PROGRESS

For the updated syllabus

Complete
Physics
for Cambridge IGCSE®

Third edition

Revision Guide

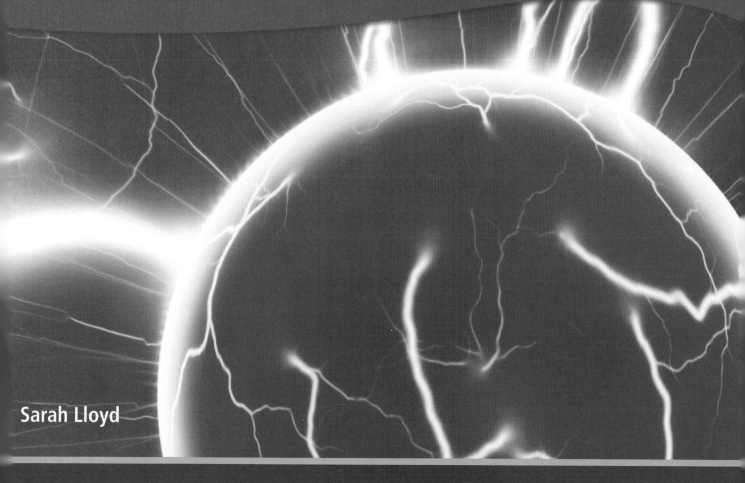

Sarah Lloyd

Oxford excellence for Cambridge IGCSE®

OXFORD

OXFORD
UNIVERSITY PRESS

Great Clarendon Street, Oxford, OX2 6DP, United Kingdom

Oxford University Press is a department of the University of Oxford. It furthers the University's objective of excellence in research, scholarship, and education by publishing worldwide. Oxford is a registered trade mark of Oxford University Press in the UK and in certain other countries

British Library Cataloguing in Publication Data
Data available

978-0-19-830874-4

10 9 8 7

Paper used in the production of this book is a natural, recyclable product made from wood grown in sustainable forests. The manufacturing process conforms to the environmental regulations of the country of origin.

Printed in India by Manipal Technologies Limited

Unless otherwise indicated, the questions and example answers were written by the author.

All past paper questions are reproduced by permission of Cambridge International Examinations.

Cambridge International Examinations bears no responsibility for the example answers to questions taken from its past question papers which are contained in this publication. The example answers, marks awarded and/or comments that appear in this book were written by the author. In examination, the way marks would be awarded to answers like these may be different.

® IGCSE is the registered trademark of Cambridge International Examinations.

Acknowledgements
The publishers would like to thank the following for permissions to use their photographs:

Cover image: Don Farrall/Getty; p38: Xuejun Li/Fotolia

Artwork by OUP and QBS Learning

Contents

How to use this revision guide

This book is designed to be used with the *Complete Physics for IGCSE* student book. It offers brief notes and simplified explanations, along with practice questions, to help you understand the physics principles required for the Cambridge IGCSE syllabus. The notes, examples, summary questions and examination questions are divided into sections that relate to the syllabus areas.

The examination questions that accompany each subsection (e.g. "Transfer of thermal energy," which is part of the "Thermal physics" section) will allow you to test yourself at regular intervals. There are also questions at the end of each section so that you can test your knowledge and understanding of the whole topic. In this way you can revise topic by topic until you have covered the entire syllabus.

The answers to all the questions in this book were written by the author.

Tips for effective revision

Active Revision for Physics

Making revision notes

Don't just write out your notes. Try to make them as brief as possible, just picking out the essential points. This is quite challenging! You could try writing your revision notes onto cue cards. They are then more portable to read on the bus/tram to school, or on the way to see somebody.

Drawing revision mind maps (spider diagrams)

This is a good way to visually summarise information. You can link ideas, which will help your understanding of the topic. Simplifying difficult concepts into diagrams will help you to reduce the information that you have to learn for the exam. Drawing out a mind map on a large sheet of paper will allow you to put in more diagrams and highlight the important points in colour. Below is an example of a simple mind map on radioactivity.

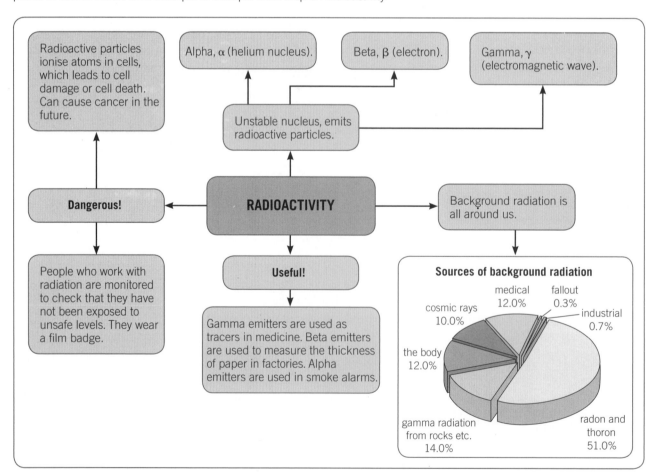

Writing your own test
Write down questions as you go through this Revision Guide. When you reach the end of the topic you can use the questions to test your knowledge and understanding and check your progress.

Getting someone else to test you
You could give one of your classmates your revision notes and ask them to test you. Or you could get them to ask you about parts of the syllabus they don't understand. This is a good test of your understanding of that topic!

Using your syllabus
You can download a copy of the syllabus from the Cambridge website. Make check lists for revision using the syllabus to indicate when you have made revision notes, practised exam questions etc. Don't forget to include a column to tick when you feel you have understood a topic.

Doing past exam questions
There are lots of examination questions throughout this book for you to try. Try revising a small part of a topic, say for about 30 minutes and then test yourself on the examination style questions provided here. When you feel ready, full past papers are available on the Cambridge website.

Making revision posters
If you make your revision notes into posters and put them in places where you will see them often, you will read them without even realising! This will help to keep topics fresh in your mind. Your family and friends will see what you have been revising and might talk to you about it, which will help the information stick in your mind.

Essential quantities and units

Quantity	Symbol	Unit
Time	t	second (s)
Force	F	newton (N)
Weight	W	newton (N)
Velocity	v	metres per second (m/s)
Speed	u	metres per second (m/s)
Distance	d or s	metre (m)
Acceleration	a	metres per second squared (m/s^2)
Mass	m	kilogram (kg)
Moment	M	newton metre (Nm)
Energy	E	joule (J)
Work done	W	joule (J)
Power	P	watt (W)
Current	I	ampere (A)
Potential difference (voltage)	V	volt (V)
Resistance	R	ohm (Ω)
Charge	Q	coulomb (C)
Frequency	f	hertz (Hz)
Pressure	P	newton per metre squared (N/m^2)
Temperature	T	degrees Celsius (°C)
Density	ρ or d	grams per centimetre cubed (g/cm^3) or kilograms per metre cubed (kg/m^3)
Wavelength	λ	metre (m)
Specific heat capacity	c	joules per kilogram degree Celsius (J/kg °C)
Specific latent heat	l	joules per kilogram (J/kg)
Momentum	p	kilogram metre per second (kg m/s)

Formulae and magic triangles

These magic triangles are a useful way of remembering how to use an equation, but are no substitute for remembering the equation itself.

To use a magic triangle, cover the quantity you are trying to find. The relationship left behind shows you how to calculate it. For example, cover up speed, u in the first triangle and you are left with $\frac{d}{t}$.

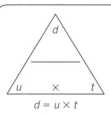

$d = u \times t$

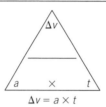

$\Delta v = a \times t$

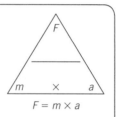

$F = m \times a$

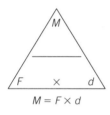

$M = F \times d$

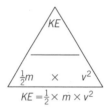

$KE = \frac{1}{2} \times m \times v^2$

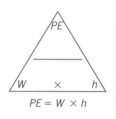

$PE = W \times h$

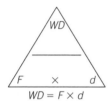

$WD = F \times d$

$F = P \times A$

$E = P \times t$

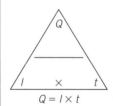

$Q = I \times t$

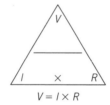

$V = I \times R$

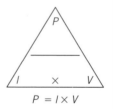

$P = I \times V$

$m = \rho \times V$

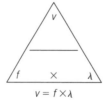

$v = f \times \lambda$

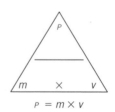

$P = m \times v$

$p = \rho g h$

$E = mc\Delta T$

$E = \Delta mL$

1 General physics

1.1 Length, volume and time

KEY IDEAS

✓ Length is measured with a ruler, tape measure, vernier calipers or micrometer screw gauge
✓ Volume may be measured with a measuring cylinder
✓ Time is measured with a stop clock or stop watch

Length measurements can be made more **precise** by using an instrument with a vernier scale such as vernier calipers or a micrometer screw gauge. A ruler measures to the nearest millimetre, vernier calipers to 1/10 mm and a micrometer 1/100 mm. Length measurements can be made more **accurate** by measuring multiples, such as the thickness of 500 sheets of paper, then divide by 500 to get the thickness of one sheet. Check the **reliability** of your measurements by repeating them. If the repeated results are similar, they are reliable. When measuring the length, l of a pendulum as in the figure (right), you need to measure to the centre of gravity. One way of doing this is to take two measurements and average them: from the fixed end of the string to the beginning of the bob, l_1 and from the fixed end of the string to the far end of the bob, l_2.

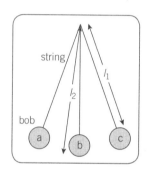

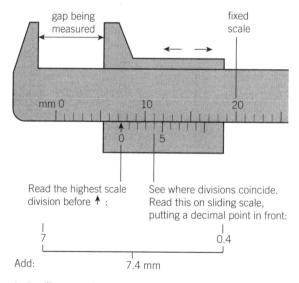

Read the highest scale division before ↑ :

7

See where divisions coincide. Read this on sliding scale, putting a decimal point in front:

0.4

Add: 7.4 mm

▲ Reading a vernier

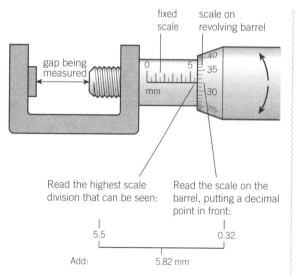

Read the highest scale division that can be seen:

5.5

Read the scale on the barrel, putting a decimal point in front:

0.32

Add: 5.82 mm

▲ Reading a micrometer

The volume of a regular solid can be found by measuring its dimensions. For example, recording the length, width and height of a cuboid (box) and multiplying these measurements together. The volume of an irregularly shaped object can also be measured using a measuring cylinder with a eureka can. Place the object in the eureka can and use the measuring cylinder to measure how much water is displaced. The volume of water displaced is equal to the volume of the object.

A digital stop watch measures time to a **precision** of 0.01 s. This is far more precise than human error will allow, which is about 0.2 s. If possible, time over as long a period as possible. When timing the time period for a simple pendulum, for example, it is more **accurate** to time 100 swings with a stop watch and then divide by the number of swings. This reduces the error due to human reaction time. Repeat time measurements to check for **reliability** of data.

▲ Eureka can and measuring cylinder

Examination style question

1. An engineering machine has a piston which is going up and down approximately 75 times per minute.

Describe carefully how a stopwatch may be used to find accurately the time for one up-and-down cycle of the piston.

Cambridge IGCSE Physics 0625 Paper 31 Q1 June 2009

1.2 Speed, velocity and acceleration

KEY IDEAS

✓ Speed = $\dfrac{\text{distance}}{\text{time}}$

✓ Acceleration = $\dfrac{\text{change in velocity}}{\text{time}}$

✓ Velocity and acceleration can have both positive and negative values

$$\text{Speed (m/s)} = \frac{\text{distance (m)}}{\text{time (s)}}$$

$$u = \frac{d}{t}$$

Worked examples

1. A car travels at a speed of 20 m/s for 30 s. How far does it travel in this time?

2. A cyclist travels 1000 m in 3 minutes. What is his speed?

3. A girl walks 3 km at 1.5 m/s. How long does her journey take?

Answers

1. $d = u \times t$
   $= 20 \times 30$
   $= 600$ m

2. $u = d/t$
   $= 1000 \div (3 \times 60)$
   $= 1000 \div 180$
   $= 5.56$ m/s

3. $t = d/u$
   $= (3 \times 1000) \div 1.5$
   $= 3000 \div 1.5$
   $= 2000$ s (33 min 20 s)

Note: In example 2, time in minutes must be converted to seconds.
   In example 3, distance in kilometres must be converted to metres.

Extended

Distance–time graphs

A journey can be represented on a graph by plotting the distance travelled on the *y*-axis and the time taken on the *x*-axis. The shape of the graph describes the journey.

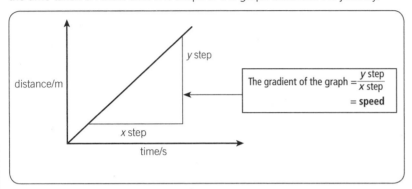

▲ Distance–time graph

Examples of distance–time graphs

1. In a crash test, a car travels at steady speed and then stops suddenly as it hits a wall.

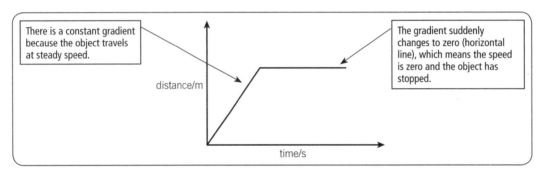

2. A runner sets off in a race, increasing her speed until she reaches her maximum speed.

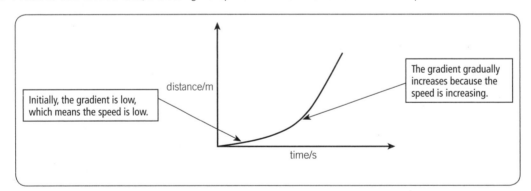

Extended

Velocity and acceleration

Velocity is a **vector** quantity; **it is equal to speed in a particular direction**. Speed is a measurement of how fast an object is moving.

Acceleration is also a vector quantity. It is equal to the **change in velocity per second**.

When an object is slowing down, if its velocity is **decreasing**, the acceleration is **negative**. We say that it is **decelerating**. See section 1.5 Forces (scalars and vectors, on page 32).

Acceleration (m/s²) = change in velocity (m/s) ÷ time (s)

$$a = \frac{v - u}{t}$$

$$a = \frac{\Delta v}{t}$$

initial velocity = u, final velocity = v,
time taken for the change in velocity = t,
change in velocity = $v - u = \Delta v$

Worked examples

1. A car increases its velocity from 10 m/s to 20 m/s in 5 s. What is its acceleration?

2. A runner has an acceleration of 10 m/s². How long does it take him to reach a speed of 5 m/s from rest? (Note 'rest' means zero velocity.)

3. A train accelerates at 9 m/s² for 5 s. If its initial velocity is 5 m/s, what is its final velocity?

Answers

1. $a = \Delta v \div t$
   $= (20 - 10) \div 5$
   $= 2$ m/s²

2. $t = \Delta v \div a$
   $= 5 \div 10$
   $= 0.5$ s

3. $\Delta v = a \times t$
   $= 9 \times 5$
   $= 45$ m/s
   $v = u + \Delta v$
   $= 5 + 45$

= 50 m/s

Speed–time graphs

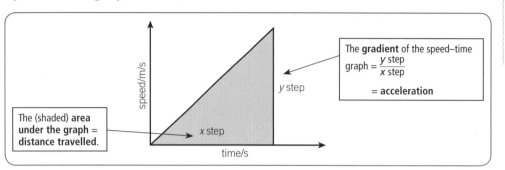

▲ Speed–time graph

The area under the graph is found by calculation, involving the units on the two axes. It is not a physical area.

Examples of speed–time graphs

1. A car accelerating until it reaches its maximum speed.

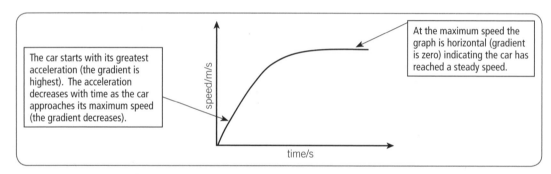

The car starts with its greatest acceleration (the gradient is highest). The acceleration decreases with time as the car approaches its maximum speed (the gradient decreases).

At the maximum speed the graph is horizontal (gradient is zero) indicating the car has reached a steady speed.

2. A runner who accelerates with constant acceleration to his maximum speed and then decelerates steadily to a stop at the end of the race.

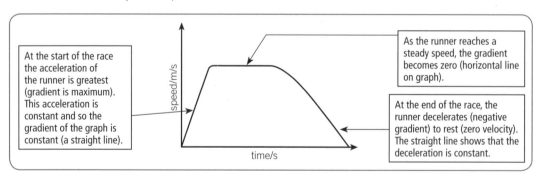

At the start of the race the acceleration of the runner is greatest (gradient is maximum). This acceleration is constant and so the gradient of the graph is constant (a straight line).

As the runner reaches a steady speed, the gradient becomes zero (horizontal line on graph).

At the end of the race, the runner decelerates (negative gradient) to rest (zero velocity). The straight line shows that the deceleration is constant.

3. A skydiver from the time she jumps from a helicopter until the moment she reaches the ground.

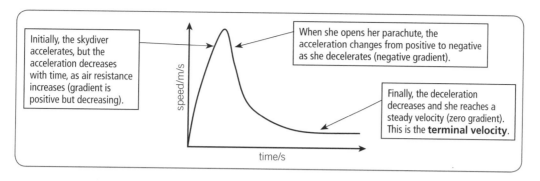

Examination style questions

1.

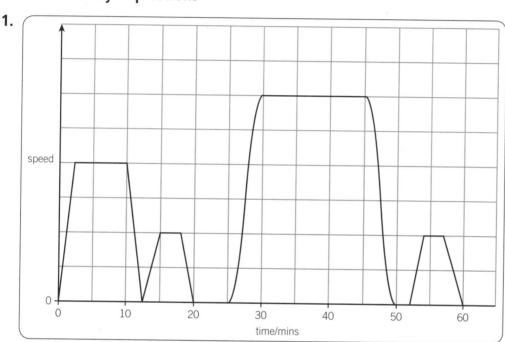

The graph above is for a 60 minute car journey.
a. Between which times is the car speed at its highest?
b. Calculate the total time for which the car is stopped.
c. State without calculation how the graph could be used
   i) to find the distance travelled in the first 12½ minutes.
   ii) to find the average speed for the journey.

2. A stone falls from the top of a building and hits the ground at a speed of 32 m/s.
The air resistance force on the stone is very small and may be neglected.
a. Calculate the time of fall.
b. On the diagram below, draw the speed–time graph for the falling stone.

Extended

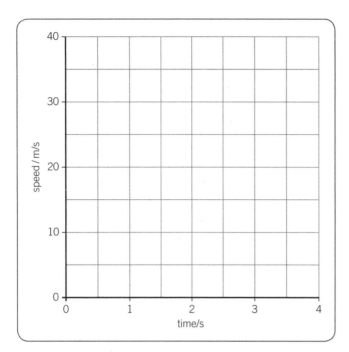

Cambridge IGCSE Physics 0625 Paper 3 Q1a November 2006

3. The figure below shows the speed–time graph for a journey travelled by a tractor.

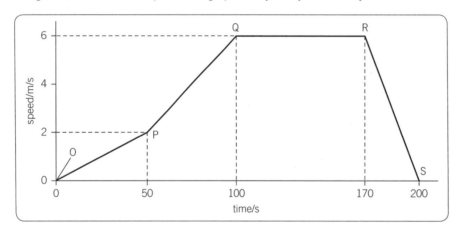

a. Use the graph to describe the motion of the tractor during the sections OP, PQ, QR and RS.

b. Which two points on the graph show when the tractor is stationary?

c. State the greatest speed reached by the tractor.

d. For how long was the tractor travelling at constant speed?

e. State how the graph may be used to find the total distance travelled during the 200 s journey. Do **not** attempt a calculation.

f. Calculate the acceleration of the tractor in section OP.

Adapted from Cambridge IGCSE Physics 0625 Paper 2 Q3 November 2006

4. Palm trees are growing every 25 m alongside the highway in a holiday resort.

The IGCSE school bus drives along the highway.

a.   It takes 2 s for the bus to travel between palm tree 1 and palm tree 2.
     Calculate the average speed of the bus between tree 1 and tree 2.

b.   It takes more than 2 s for the bus to travel from tree 2 to tree 3.
     State what this information indicates about the speed of the bus.

c.   The speed of the bus continues to do what you have said in **(b)**. State how the
     time taken to go from tree 3 to tree 4 compares with the time in **(b)**.

Cambridge IGCSE Physics 0625 Paper 2 Q2 November 2005

5. In a training session, a racing cyclist's journey is in three stages.
Stage 1   He accelerates uniformly from rest to 12 m/s in 20 s.
Stage 2   He cycles at 12 m/s for a distance of 4800 m.
Stage 3   He decelerates uniformly to rest.
The whole journey takes 500 s.

a.   Calculate the time taken for stage 2.

b.   Copy the grid below and draw a speed–time graph of the cyclist's ride.

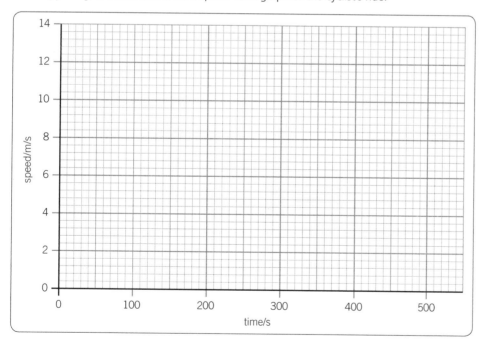

c.   Show that the total distance travelled by the cyclist is 5400 m.

d.   Calculate the average speed of the cyclist.

e.   Calculate the cyclist's acceleration in stage 1.

Adapted from Cambridge IGCSE Physics 0625 Paper 2 Q2 June 2007

1.3 Mass and weight

KEY IDEAS

✓ Mass is a quantity related to the inertia of an object, measured in kg
✓ Weight is the force, in N, on a mass due to a gravitational field
✓ On Earth, the gravitational field strength is 10 N/kg

Mass is the amount of **matter** that makes up an object. It is measured in kilograms (**kg**). All masses have a quality called "inertia", the tendency to keep moving if already moving and stay still if already still.

For example, a car in a crash test: when the car hits the wall, it decelerates to rest in a short time. The crash test dummy has inertia due to its mass and so it keeps moving forwards at the same speed as before the car hit the wall. Although it looks as though the dummy has been thrown forward, there is no net forward force on it.

Weight is the force on a mass due to gravity. It is measured in newtons (**N**).

weight (N) = mass (kg) × gravitational field strength (N/kg)

$$W \quad = \quad m \quad \times \quad g$$

On Earth, the gravitational field strength, g = **10 N/kg**. This is also called the acceleration due to gravity or the acceleration of freefall and has an alternative unit of m/s^2.

Examination style question

Some IGCSE students were asked to write statements about mass and weight.

Their statements are printed below. Choose the **two** correct statements.

Mass and weight are the same thing.

Mass is measured in kilograms.

Weight is a type of force.

Weight is the acceleration caused by gravity.

Cambridge IGCSE Physics 0625 Paper 2 Q2 November 2006

1.4 Density

KEY IDEAS

✓ Density $= \dfrac{\text{mass}}{\text{volume}}$, measured in g/cm³ or kg/m³

Density is a quantity related to how closely packed the particles in a material are, as well as the mass of the particles.

$$\text{density (g/cm}^3\text{)} = \frac{\text{mass (g)}}{\text{volume (cm}^3\text{)}}$$
$$\rho = \frac{m}{V}$$

A simple method of measuring the density of an object (if its density is greater than that of water):

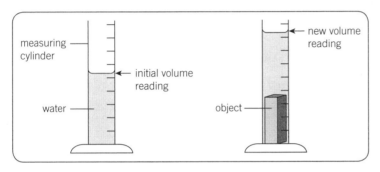

- Find the mass of the object using a balance.
- Approximately half fill a measuring cylinder with water. Read the volume of water from the measuring cylinder scale.
- Place the object in the measuring cylinder and take the new volume reading.
- Calculate (new volume reading – initial volume reading) to find the volume of the object.
- Calculate the density by dividing the mass by the volume.

Worked examples

1. The mass of a stone is found on a top pan balance. It has a mass of 120.02 g. A measuring cylinder is filled to a volume of 60 cm³ with water. When the stone is placed in the measuring cylinder, the new water level is 95 cm³. Find the density of the stone.

2. An object of known density of 2.7 g/cm³ is placed into a measuring cylinder of water. The level of the water rises from 45 cm³ to 72 cm³. What is the mass of the object?

3. A metal block of density 3.2 g/cm³ and mass 90 g is placed in a measuring cylinder containing 65 cm³ of water. What is the new water level?

Answers

1. Volume = 95 – 60 = 35 cm³

$\rho = \dfrac{m}{V} = \dfrac{120.02}{35} = 3.4$ g/cm³

2. Volume = 72 – 45 = 27 cm³

$m = V \times \rho = 27 \times 2.7 = 73$ g

3. $V = \dfrac{m}{\rho} = \dfrac{90}{3.2} = 28$ cm³

New water level = 65 + 28 = 93 cm³

Predicting whether an object will sink or float

An object will **sink** if its density is greater than the density of the liquid in which it is placed. If the object has the same density as the liquid or less, it will **float**. For example:

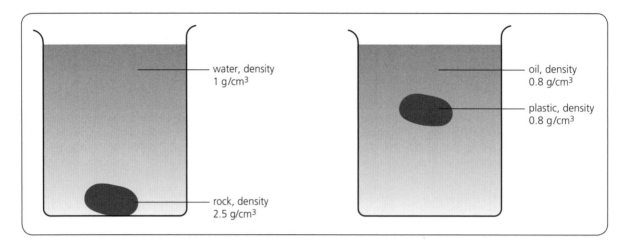

Examination style questions

1. A student is given a spring balance with a newton scale. She is told that the acceleration due to gravity is 10 m/s^2.
Describe how she could find the mass of a toy car.
Describe how she could go on to find the average density of the toy car.

Adapted from Cambridge IGCSE Physics 0625 Paper 3 Q1b November 2005

2. A student used a suitable measuring cylinder and a spring balance to find the density of a sample of a stone.
   a.   Describe how the measuring cylinder is used, and state the readings that are taken.
   b.   Describe how the spring balance is used, and state the reading that is taken.
   c.   Write down an equation from which the density of the stone is calculated.
   d.   The student then wishes to find the density of cork. Suggest how the apparatus and the method would need to be changed.

Cambridge IGCSE Physics 0625 Paper 3 Q1b November 2006

3. Fig. (a) shows a measuring cylinder, containing some water, on a balance.
Fig. (b) shows the same arrangement with a stone added to the water.

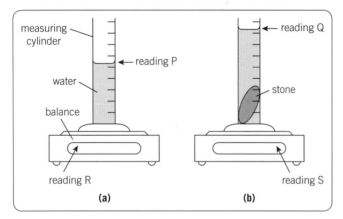

   a.   Which two readings should be subtracted to give the volume of the stone?
   b.   Which two readings should be subtracted to give the mass of the stone?

c. In a certain experiment,

mass of stone = 57.5 g,

volume of stone = 25 cm³.

i)   Write down the equation linking density, mass and volume.

ii) Calculate the density of the stone.

Cambridge IGCSE Physics 0625 Paper 2 Q3 June 2006

Practical question

An IGCSE student is determining the density of a metal alloy.

The student is provided with several metal rods, as shown on the right.

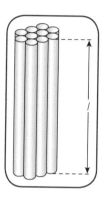

1. Measure with a ruler the length, l, of one of the rods.

2. The student measured the diameter of one of the rods with a ruler and found it to be 0.6 cm. Calculate the cross-sectional area, A, of the rod.

3. Use this value to calculate the volume, V, of one rod and hence the whole bundle.

The student used a balance to find the mass of the bundle and found it to be 59.1 g. Calculate the density of the metal alloy.

1.5 Forces

Forces can produce a **change in size or shape of an object**. For example, loading a spring will increase its length.

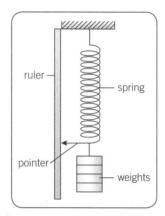

An experiment to find how the extension of a spring varies with the force applied

* Measure the original position of the spring using the pointer and ruler, to the nearest mm.
* Add a 100 g mass hanger and measure the new position.
* Repeat 6 times, adding a 100 g mass each time.
* Calculate the extension by subtracting the original position from each subsequent position reading.
* Plot a graph of extension against force, where force = mass × 10 N/kg.

Graph of results

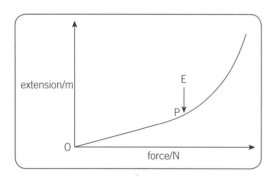

Conclusion
From **0** to **P**, extension is directly proportional to the force applied.
Beyond P, the extensions are larger for the same increase in force.

At **E**, the elastic limit is reached. Beyond this point, the spring will not return to its original length when the force is removed.

Hooke's Law

If a material obeys Hooke's Law, the extension is directly proportional to the applied force, provided that the elastic limit is not exceeded.

$$F = kx$$

Where F = applied force (N), k = force constant for object under test (N/m), x = extension (m)

Force, mass and acceleration

A force can **accelerate** an object. The larger the force on the object, the greater the acceleration if the mass stays constant. The larger the mass of the object, the smaller the acceleration if the force stays constant.

Force (N) = mass (kg) × acceleration (m/s²)

$$F = m \times a$$

Worked examples

1. A force of 10 N acts on an object of mass 5 kg. What is the acceleration of the object?

2. A force of 15 N causes an object to accelerate at 2 m/s². What is its mass?

3. A mass of 3 kg has a deceleration of 5 m/s². What force acts on it?

Answers

1. $a = F \div m$
$= \frac{10}{5}$
$= 2$ m/s²

2. $m = F \div a$
$= \frac{15}{2}$
$= 7.5$ kg

3. $F = m \times a$
$= 3 \times -5$
$= -15$ N

Note: In example 3, the object is decelerating so acceleration is negative.

Friction and air resistance

Frictional forces act between two surfaces and always oppose the direction of motion of an object. Where friction acts, work is done, which transforms the kinetic energy of the object into **heat**. **Air resistance** is a form of friction, which acts to oppose the motion of an object as it moves through air.

Resultant force

The **resultant** force acting on an object is the **net** or **overall** force when the **size** and **direction** of all the forces acting are taken into account. **Force** is a **vector** quantity.

If the resultant force of an object is zero, the object remains at rest or continues at constant speed in a straight line.

Examples

1.

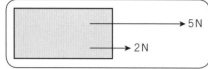

Resultant force = 5 N + 2 N = 7 N

The forces are **added** together because they act in the **same** direction.

2.

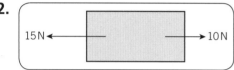

Resultant force = 15 N − 10 N
= 5 N (to the left)

The forces are **subtracted** because they act in **opposite** directions.

A force can cause an object to **change direction**. The object will move in a circle if the force acts perpendicular (at a right angle) to the direction that the object is travelling.

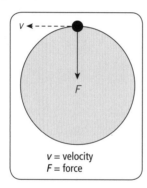

v = velocity
F = force

The force acts at **right angles** to the direction that the object is moving. The force does not do any work on the object because the object does not move in the direction of the force. The force constantly changes the direction of the object which means its **velocity changes**, but its **speed stays constant**. The object accelerates towards the centre of the circle.

The force which causes an object to move in a circle is called the **centripetal force**. The centripetal force increases if:

- the mass of the object increases
- the speed of the object increases
- the radius of the circle decreases.

If the force which is providing the centripetal acceleration is suddenly removed, the object will move on a tangent to the original circle.

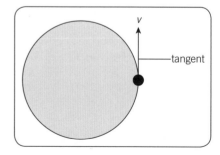

Examination style questions

1. The length of a spring is measured when various loads from 1.0 N to 6.0 N are hanging from it. The figure below gives a graph of the results.

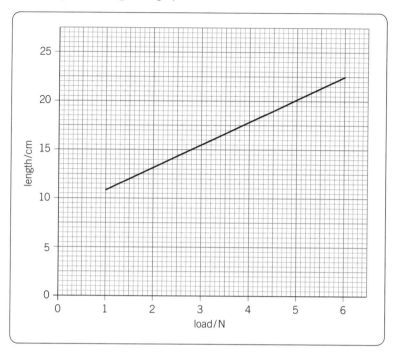

Use the graph to find:
a. the length of the spring with no load
b. the length of the spring with 4.5 N attached
c. the extension caused by a 4.5 N load.

Cambridge IGCSE Physics 0625 Paper 2 Q1 November 2005

2. A mass of 3.0 kg accelerates at 2.0 m/s² in a straight line.
a. State why the velocity and the acceleration are both described as vector quantities.
b. Calculate the force required to accelerate the mass.

Cambridge IGCSE Physics 0625 Paper 3 Q3a&b June 2005

3. In an experiment, forces are applied to a spring as shown in (a). The results of this experiment are shown in (b).

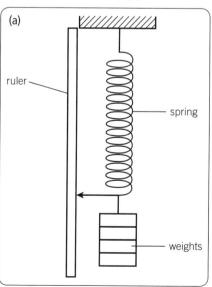

Extended

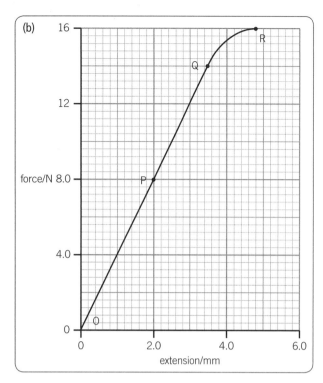

a. What is the name given to the point marked Q in (b)?

b. For the part OP of the graph, the spring obeys Hooke's Law.
State what this means.

c. The spring is stretched until the force and extension are shown by the point R on the graph. Compare how the spring stretches, as shown by the part of the graph OQ, with that shown by QR.

d. The part OP of the graph shows the spring stretching according to the expression
$$F = kx.$$
Use values from the graph to calculate the value of k.

Cambridge IGCSE Physics 0625 Paper 3 Q2 November 2006

4. The figure below shows the speed-time graphs for two falling balls.

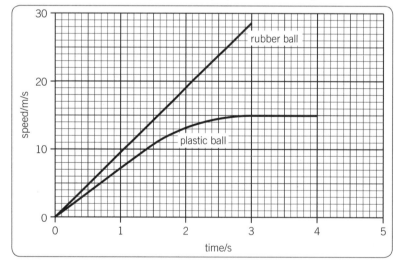

Both balls fall from the same height above the ground.

a. Use the graphs to find
   i) the average acceleration of the falling rubber ball during the first 3.0 s,
   ii) the distance fallen by the rubber ball during the first 3.0 s,
   iii) the terminal velocity of the plastic ball.
b. Both balls have the same mass but the volume of the plastic ball is much greater than that of the rubber ball. Explain, in terms of the forces acting on each ball, why the plastic ball reaches a terminal velocity, but the rubber ball does not.
c. The rubber ball has a mass of 50 g. Calculate the gravitational force acting on the rubber ball.

Cambridge IGCSE Physics 0625 Paper 31 Q1 June 2008

5. The points plotted on the grid below were obtained from a spring-stretching experiment.

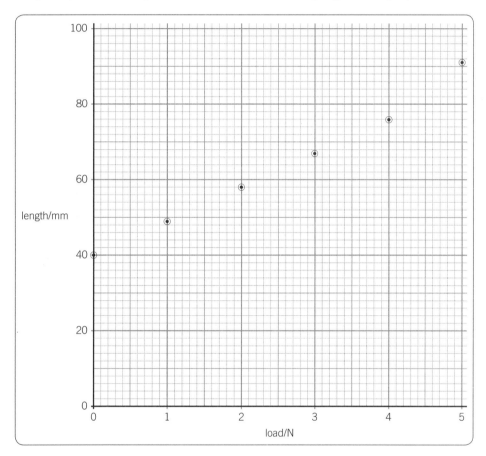

a. Using a straight edge, draw a straight line through the first 5 points. Extend your line to the edge of the grid.
b. Suggest a reason why the sixth point does not lie on the line you have drawn.
c. Calculate the extension caused by the 3 N load.
d. A small object is hung on the unloaded spring, and the length of the spring becomes 62 mm.
Use the graph to find the weight of the object.

Cambridge IGCSE Physics 0625 Paper 2 Q9 November 2006

Practical question

An IGCSE class is investigating the effect of a load on a metre rule attached to a spring. The apparatus is shown in the diagram below.

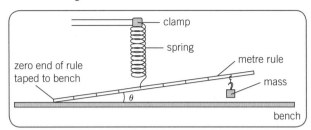

The zero end of the metre rule is taped to the bench to stop it slipping. The spring is attached to the rule at the 40.0 cm mark and the masses are attached at the 90.0 cm mark. The masses are added 10 g at a time and the angle, θ, between the bench and the rule measured with a protractor.

One student's results are shown below.

m	θ
0	29
10	28
20	26
30	25
40	22
50	19

1. Complete the column headings.

2. One student suggests that m and θ should be directly proportional to each other. Plot a graph of θ (y-axis) against m (x-axis). Using your graph show whether this prediction is correct. State your reason.

Adapted from Cambridge IGCSE Physics 0625 Paper 6 Q1a & b November 2005

Turning effect and equilibrium

The turning effect or **moment** of a force about a pivot is equal to the force multiplied by its perpendicular distance from the pivot.

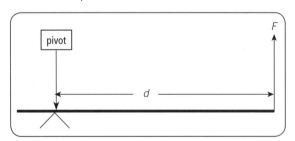

Moment (Nm) = Force (N) × distance (m)

$$M \quad = \quad F \quad \times \quad d$$

If an object is in equilibrium there is no resultant turning effect and no resultant force.

Worked example

1. A force of 2.0 N acts at distance of 3.0 m from a pivot. Find the moment of the force.

2. A force of 5.0 N provides a moment of 15 Nm about a pivot. What is the distance of the force from the pivot?

3. A force provides a moment of 20 Nm about a pivot at a distance of 2.0 m. What is the
size of the force?

Answers

1. $M = F \times d$

$= 2.0 \times 3.0$

$= 6.0$ Nm

2. $d = \dfrac{M}{F}$

$= \dfrac{15}{5.0}$

$= 3.0$ m

3. $F = \dfrac{M}{d}$

$= \dfrac{20}{2.0}$

$= 10$ N

Examples of objects in equilibrium

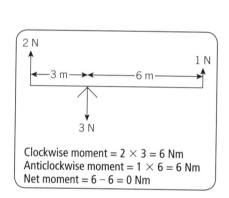

Clockwise moment = $2 \times 3 = 6$ Nm
Anticlockwise moment = $1 \times 6 = 6$ Nm
Net moment = $6 - 6 = 0$ Nm

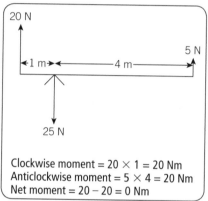

Clockwise moment = $20 \times 1 = 20$ Nm
Anticlockwise moment = $5 \times 4 = 20$ Nm
Net moment = $20 - 20 = 0$ Nm

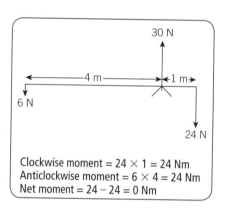

Clockwise moment = $24 \times 1 = 24$ Nm
Anticlockwise moment = $6 \times 4 = 24$ Nm
Net moment = $24 - 24 = 0$ Nm

Taking moments about the pivot in each case.

An experiment to show that there is no net moment on an object in equilibrium

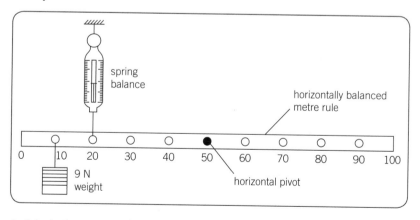

Anticlockwise moment due to the 9 N weight = $9 \times 0.4 = 3.6$ Nm
Reading on the spring balance = 12 N
Moment due to the force of the spring balance = $12 \times 0.3 = 3.6$ Nm

Conclusion: in equilibrium, clockwise moment = anticlockwise moment

Examination style questions

1. a. State the two factors on which the turning effect of a force depends.
   b. Forces F_1 and F_2 are applied vertically downwards at the ends of a beam resting on a pivot P. The beam has weight W. The beam is shown in the diagram below.

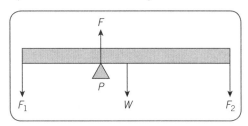

      i) Copy and complete the statements about the two requirements for the beam to be in equilibrium.

          1. There must be no resultant ....

          2. There must be no resultant ....

      ii) The beam is in equilibrium. F is the force exerted on the beam by the pivot P. Complete the following equation about the forces on the beam.

         $F =$ ....

      iii) Which one of the four forces on the beam does **not** exert a moment about P?

   Cambridge IGCSE Physics 0625 Paper 2 Q5 November 2006

2. The diagram below shows apparatus for investigating moments of forces.

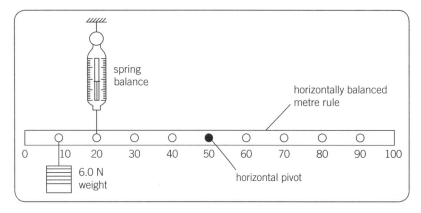

The uniform metre rule shown is in equilibrium.
   a. Write down two conditions for the metre rule to be in equilibrium.
   b. Show that the value of the reading on the spring balance is 8.0 N.
   c. The weight of the uniform metre rule is 1.5 N. Calculate the force exerted by the pivot on the metre rule and state its direction.

   Cambridge IGCSE Physics 0625 Paper 3 Q2 November 2005

Practical question

The IGCSE class is determining the weight of a metre rule.

Below is a diagram of the apparatus.

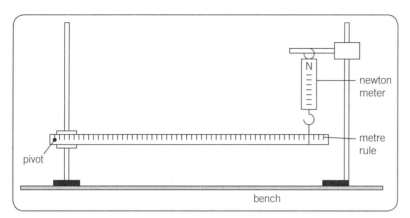

A metre rule is supported at one end by a pivot through the 1.0 cm mark. The other end is supported at the 91.0 cm mark by a newton meter hanging from a clamp.

1. Describe how you would check that the metre rule is horizontal. You may draw a diagram if you wish.

2. The students record the force F shown on the newton meter and the distance d from the pivot to the 91 cm mark. They then repeat the experiment several times using a range of values of the distance d. The readings are shown in the table.

F/N	d/m	$\frac{1}{d}$ / $\frac{1}{m}$
0.74	0.900	
0.78	0.850	
0.81	0.800	
0.86	0.750	
0.92	0.700	

Copy the table. Calculate and record on your table the values of $\frac{1}{d}$.

3. a. On graph paper, plot a graph of F/N (y-axis) against $\frac{1}{d}$/$\frac{1}{m}$ (x-axis). Start the y-axis at 0.7 and the x-axis at 1.0.
   b. Draw the line of best fit on your graph.
   c. Determine the gradient G of the line.

4. Calculate the weight of the metre rule using the equation

$$W = \frac{G}{k} \quad \text{where } k = 0.490 \text{ m.}$$

Cambridge IGCSE Physics 0625 Paper 6 Q5 June 2006

Centre of mass

The centre of mass of an object is the point on the object where the mass can be considered to be concentrated and hence where the weight of the object can be considered to act.

The centre of mass of a very thin object (a lamina) can be found by experiment:

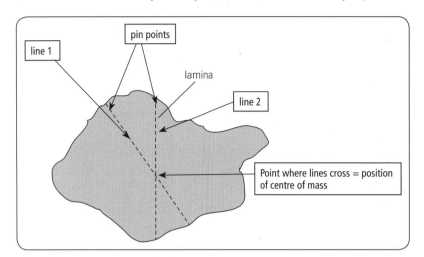

Push a pin through a point on the edge of the lamina and allow it to swing freely. Use a plumb line (a small mass on a piece of string) to mark a vertical line from the pin point across the lamina. Repeat for a second point on the edge of the lamina. Where the two lines cross is the position of the centre of mass.

The position of the centre of mass affects the **stability** of an object.

For example:

If object 1 is tilted through a small angle, the **weight will act outside the base**. There will be a net moment on object 1 that will cause it to fall over. If object 2 is tilted through a small angle the weight will **still act inside the base**. There will be a net moment on object 2 that will cause it to go back to its original position and it will not fall over.

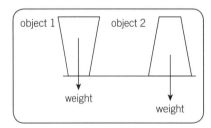

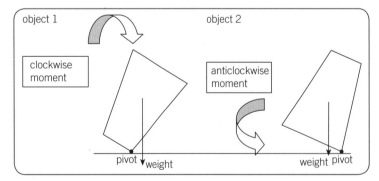

Examination style question

1. A piece of stiff cardboard is stuck to a plank of wood by means of two sticky-tape "hinges". This is shown below.

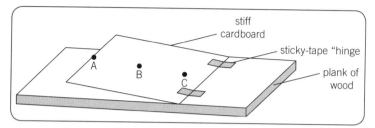

a. Initially, the cardboard is flat on the plank of wood. A box of matches is placed on it. The cardboard is then slowly raised at the left hand edge, as shown below. State the condition for the box of matches to fall over.

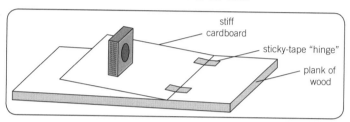

b. The box of matches is opened, as shown below. The procedure in (a) is repeated.

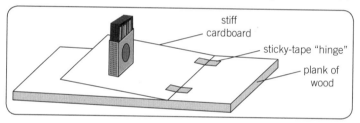

   i) Copy and complete the sentence below, using either the words "greater than" or "the same as" or "less than".

   In (b), the angle through which the cardboard can be lifted before the box of matches falls is .... the angle before the box of matches falls in (a).

   ii) Give a reason for your answer to (i).

Cambridge IGCSE Physics 0625 Paper 2 Q3b&c June 2007

Extended

Scalars and vectors

A **scalar** quantity has **size** only.

A **vector** quantity has **size** and **direction**.

Examples of scalar quantities	Examples of vector quantities
Mass	Velocity
Energy	Acceleration
Time	Force

Resultants

To calculate the resultant (overall) force on a point acted on by two forces, F_1 and F_2 you can draw a scale diagram.

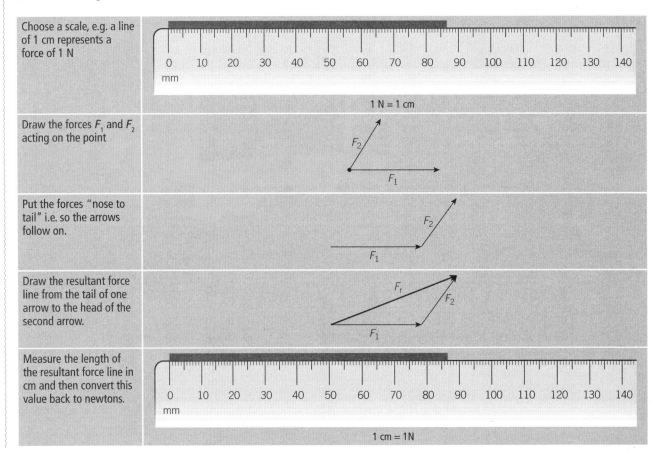

Choose a scale, e.g. a line of 1 cm represents a force of 1 N	1 N = 1 cm
Draw the forces F_1 and F_2 acting on the point	
Put the forces "nose to tail" i.e. so the arrows follow on.	
Draw the resultant force line from the tail of one arrow to the head of the second arrow.	
Measure the length of the resultant force line in cm and then convert this value back to newtons.	1 cm = 1N

Extended

Examination style question

1. a. In an accident, a truck goes off the road and into a ditch. Two breakdown vehicles A and B are used to pull the truck out of the ditch, as shown below.

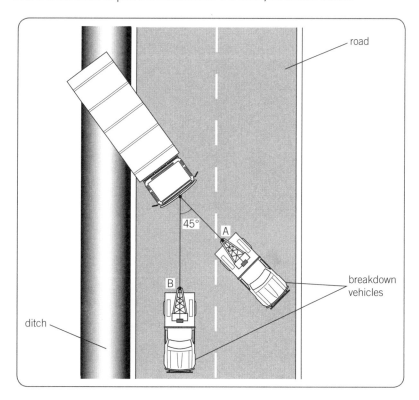

At one point in the rescue operation, breakdown vehicle A is exerting a force of 4000 N and breakdown vehicle B is exerting a force of 2000 N.

   i) Using a scale of 1 cm = 500 N, make a scale drawing to show the resultant force on the truck.

   ii) Use your diagram to find the magnitude and direction of the resultant force on the truck.

b. i) State why the resultant force is an example of a vector quantity.

   ii) Give an example of a vector quantity that is not a force.

Cambridge IGCSE Physics 0625 Paper 31 Q4 June 2009

1.6 Momentum

Momentum is a vector quantity linked to mass and velocity by the equation:

momentum = mass × velocity

(kg m/s)   (kg)   (m/s)

Worked examples

1. What is the momentum of a 1000 kg car moving at 15 m/s?

2. What is the velocity of a runner of mass 60 kg with a momentum of 300 kg m/s?

3. What is the mass of a bicycle with momentum 50 kg m/s moving at 5 m/s?

Answers

1. $p = m \times v$

$= 1000 \times 15$

$= 15\,000$ kg m/s

2. $v = p \div m$

$= \dfrac{300}{60}$

$= 5$ m/s

3. $m = p \div v$

$= \dfrac{50}{5}$

$= 10$ kg

Impulse

Impulse = change in momentum = $mv - mu$

(where v = final velocity and u = initial velocity)

Worked examples

1. A trolley of mass 0.5 kg, moving with velocity 2 m/s increases its velocity to 3 m/s. What was the impulse on the trolley?

2. A toy car of mass 0.1 kg is moving at a velocity of 0.5 m/s when it receives an impulse of 0.2 kg m/s. What is its velocity?

3. After receiving an impulse of 15 kg m/s, a shopping trolley of mass 30 kg has a velocity of 1.5 m/s. What was the initial velocity of the shopping trolley?

Answers

1. Final momentum = $0.5 \times 3 = 1.5$ kg m/s

Initial momentum = $0.5 \times 2 = 1$ kg m/s

Impulse = change in momentum = $1.5 - 1 = 0.5$ kg m/s

2. Initial momentum = $0.1 \times 0.5 = 0.05$ kg m/s

Final momentum = impulse + initial momentum

$= 0.2 + 0.05 = 0.25$ kg m/s

Final velocity = final momentum ÷ mass = $\dfrac{0.25}{0.1} = 2.5$ m/s

3. Final momentum = $30 \times 1.5 = 45$ kg m/s

initial momentum = final momentum − impulse

$= 45 - 15 = 30$ kg m/s

Initial velocity = initial momentum ÷ mass = $\dfrac{30}{30} = 1$ m/s

> **Newton's second law** states that force is equal to change in momentum per second.
> It follows that: **impulse = force × time**.

Principle of Conservation of Momentum

> The **momentum before** a collision is equal to the **momentum after** provided that no
> external forces act.

Worked examples

1. A car of mass 1000 kg travelling at a velocity of 25 m/s collides with another car, of mass 1500 kg, which is at rest. The two cars stick and move off together. What is the velocity of the two cars after the collision?

2. A boy of mass 50 kg, running at a speed of 4 m/s, jumps onto a skateboard of mass 2 kg which is already moving in the same direction as the boy at 2 m/s. What is the velocity of the boy and the skateboard after the boy jumps on?

3. A ball of mass 0.5 kg moving at a velocity of 5 m/s collides with a ball of mass 0.2 kg moving in the opposite direction at 2 m/s. If the 0.5 kg ball rebounds after the collision at 1 m/s, calculate the size and direction of the velocity of the 0.2 kg ball.

Answers

1.

$V = 0$ $V = 25$ m/s $V = ?$

Before the collision   After the collision

Momentum before collision = $(1500 \times 0) + (1000 \times 25) = 25\,000$ kg m/s

Momentum after collision = $2500 \times v$

Momentum before = momentum after

$25\,000 = 2500 \times v$

$v = \dfrac{25\,000}{2500}$

$= 10$ m/s

Extended

Extended

2. $v = 4$ m/s $v = 2$ m/s $v = ?$

Before

After

Momentum before $= (50 \times 4) + (2 \times 2) = 204$ kg m/s

Momentum after $= 52 \times v$

Momentum before = momentum after

$$204 = 52 \times v$$

$$v = \frac{204}{52}$$

$$= 3.9 \text{ m/s}$$

3. $v = 5$ m/s $v = 2$ m/s $v = 1$ m/s $v = ?$

Momentum before $= (0.5 \times 5) - (0.2 \times 2) = 2.1$ kg m/s

Momentum after $= -(0.5 \times 1) + (0.2 \times v)$

Momentum before = momentum after

$$2.1 = -(0.5 \times 1) + (0.2 \times v)$$

$$0.2v = 2.6$$

$$v = 13 \text{ m/s}$$

in the same direction as the 0.5 kg ball was travelling before the collision

Worked examples of explosions

1. A bullet of mass 20 g leaves a gun of mass 600 g at a velocity of 200 m/s. What is the recoil velocity of the gun?

2. A cannon ball of mass 3 kg is fired at a velocity of 50 m/s. The cannon recoils at 2.5 m/s. What is the mass of the cannon?

Answers

1. $v = 0$ $v = 200$ m/s $v = ?$

Before   After

Momentum before $= 0$ kg m/s

Momentum after $= (0.02 \times 200) - (0.6 \times v)$

Extended

Note: The minus sign indicates the gun moves in the opposite direction to the bullet.

Momentum before = momentum after

$(0.02 \times 200) - (0.6 \times v) = 0$

$$v = \frac{(0.02 \times 200)}{0.6}$$

$$v = 6.7 \text{ m/s}$$

2.

$v = 0$  $v = 50$ m/s $v = -2.5$ m/s

Momentum before = 0 kg m/s

Momentum after = $(3 \times 50) - (m \times 2.5)$

Note: The minus sign indicates the cannon moves in the opposite direction to the bullet.

Momentum before = momentum after

$(3 \times 50) - (m \times 2.5) = 0$

$$m = \frac{3 \times 50}{2.5}$$

$$m = 60 \text{ kg}$$

1.7 Energy, work and power

Energy

An object may have energy because it is moving or because of its position. Energy can be **transferred** from one place to another, **transformed** from one type to another or **stored**. The unit of energy is the **joule (J)**.

Types of energy

Gravitational potential	The energy gained as an object is moved away from the Earth, e.g. a book being lifted onto a shelf
Kinetic	The energy an object has due to its movement, e.g. a person running
Chemical	Stored energy that can be released in a chemical reaction, e.g. a battery, fuel such as coal
Strain	The energy stored when an object changes shape, e.g. a stretched rubber band
Electric	The energy carried by an electric current
Sound	The energy carried by a sound wave
Internal energy	The total kinetic and potential energies of all of the particles in an object
Thermal (heat) energy	The energy released when the temperature of a hot object decreases due to a decrease in its internal energy
Nuclear	Stored energy that can be released in a nuclear reaction, e.g. energy stored in the Sun
Light	Energy given off, for example, by very hot objects

Energy is transferred during events and processes. For example:

- Heat energy can be **transferred** from a hot object to a cooler one.
- Kinetic energy can be **transferred** from one car to another in a collision.
- Electrical energy can be **transferred** in a circuit.

Energy transformations

In an energy transformation, energy is converted from one type to another. For example:

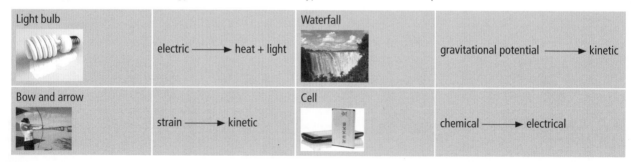

Light bulb		Waterfall	
	electric ⟶ heat + light		gravitational potential ⟶ kinetic
Bow and arrow		Cell	
	strain ⟶ kinetic		chemical ⟶ electrical

Extended

Principle of conservation of energy

Energy cannot be created or destroyed. It is transformed from one form to another.

Examples

1.

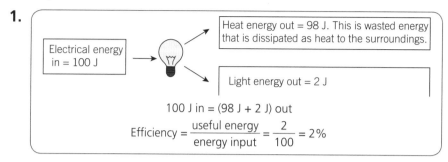

Electrical energy in = 100 J

Heat energy out = 98 J. This is wasted energy that is dissipated as heat to the surroundings.

Light energy out = 2 J

100 J in = (98 J + 2 J) out

$$\text{Efficiency} = \frac{\text{useful energy}}{\text{energy input}} = \frac{2}{100} = 2\%$$

2.

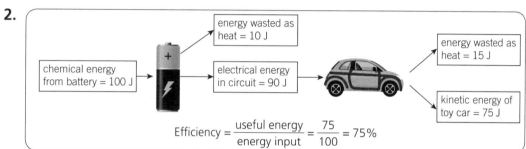

energy wasted as heat = 10 J

chemical energy from battery = 100 J

electrical energy in circuit = 90 J

energy wasted as heat = 15 J

kinetic energy of toy car = 75 J

$$\text{Efficiency} = \frac{\text{useful energy}}{\text{energy input}} = \frac{75}{100} = 75\%$$

Examination style questions

1. a. The principle of conservation of energy states that energy can neither be created nor destroyed.
What, then, *does* happen to the energy supplied to a device such as a motor or a television?

b. The television shown in the figure (right) is switched on to watch a programme. During this time, 720 kJ of electrical energy is supplied.

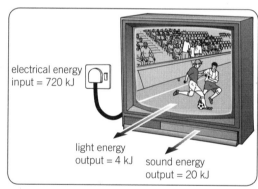

electrical energy input = 720 kJ

light energy output = 4 kJ   sound energy output = 20 kJ

   i) From the information in the figure, find the total energy provided for the viewer to see and hear the television during this programme.

   ii) Suggest what happens to the rest of the energy supplied.

   iii) Calculate how much energy is involved in **(b)(ii)**.

   iv) Comment on the efficiency of the television.

Cambridge IGCSE Physics 0625 Paper 22 Q5 June 2012

2. Two geologists are collecting rocks from the bottom of a cliff. The rocks are loaded into a basket and then pulled up the cliff on the end of a rope, as shown in the figure. The basket of rocks is brought to rest at the top of the cliff.

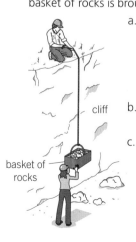

cliff

basket of rocks

a.   i) Which form of energy that the basket possesses is significantly greater at the top of the cliff than when it is at the bottom of the cliff?

   ii) Which two measurements must be made in order to calculate the increase in energy in **(i)**?

b. Which form of energy in his body has the man at the top of the cliff used in order to raise the basket of rocks?

c. State the measurement needed, in addition to those in **(a)(ii)**, in order to calculate the useful power developed by the man at the top of the cliff.

Cambridge IGCSE Physics 0625 Paper 22 Q4a&b June 2012

Energy resources

Energy resources are used to produce electrical energy from other forms of energy.

Coal, oil and gas fired power stations	The **chemical** energy in the fuel is released by burning. The chemical energy is transformed to heat energy which is used to heat water and increase its **internal** energy, turning it into steam. The steam turns turbines, transferring its **kinetic** energy to them. The kinetic energy is transformed to **electrical** energy in the generator.
Geothermal power stations	Water is pumped underground and gains **heat** energy from the hot rocks deep underground. The heat energy is then converted to **kinetic** energy in the turbines, which turn the generator to produce **electrical** energy.
Hydroelectric power stations	The **gravitational potential** energy of the falling water is transformed to **kinetic** energy as the water passes through the turbines. The turbines turn the generator to produce **electrical** energy.
Solar cells	The **light** energy from the Sun is transformed to **electrical** energy in the solar cell.
Solar power station	The **heat** energy from the Sun is concentrated by a series of curved mirrors, which focus the energy into one place. This heat energy converts water to steam, which turns the turbines, giving them **kinetic** energy. The kinetic energy is transformed to **electrical** energy in the generator.
Wave power	As the turbines bob on the surface of the sea, **gravitational potential** energy is transformed to **kinetic** energy. The kinetic energy is then transformed into **electrical** energy in the generators.
Tidal power	As the tide comes in, the water builds up behind the dam and gains **gravitational potential** energy. When the water is released, the gravitational potential energy is transformed to **kinetic** energy and then to **electrical** energy in the generator.
Nuclear power	The **nuclear** energy stored in uranium-235 is released when the uranium nuclei split in a process called nuclear fission. The nuclear energy is transformed to **heat** energy which is used to turn water to steam. The steam turns the turbines and then this **kinetic** energy is transformed to **electrical** energy in the generator.
Wind power	The **kinetic** energy of moving air is transferred to the blades, which spin, turning the generator. In the generator **kinetic** energy is transformed into **electrical**.

Advantages and disadvantages of energy resources

Energy Resources	Advantages	Disadvantages
Fossil fuel	Coal will not run out for many years Reliable source of energy	Produces gaseous pollution such as carbon dioxide (which contributes to global warming) and sulfur dioxide (which causes acid rain) Non-renewable (will run out eventually)
Geothermal	Renewable (will not run out) Does not produce carbon dioxide or sulphur dioxide Reliable source of energy	There are few areas of the world where geothermal energy is available
Hydroelectric	Renewable Does not produce carbon dioxide or sulfur dioxide Energy can be stored as gravitational potential energy of water, to be used later	Land is flooded to build the reservoirs, which destroys farmland, habitats and sometimes people's homes
Solar cells	Can be used remotely in areas that do not have a National Grid Does not produce carbon dioxide or sulfur dioxide Renewable Low maintenance	Solar cells are relatively inefficient and expensive to produce, so the payback time can be long Unreliable: cannot be used at night and output in cloudy weather is low
Solar power station	Renewable Does not produce carbon dioxide or sulfur dioxide	The arrays of mirrors occupy a large area Unreliable: cannot be used at night and output in cloudy weather is low
Wave	Renewable Does not produce carbon dioxide or sulfur dioxide	Unreliable: cannot be used in stormy weather conditions and output is low on a calm day Maintenance is difficult Can interfere with shipping
Tidal	Renewable Does not produce carbon dioxide or sulfur dioxide Reliable: there are two tides per day	Expensive to build There are few places with suitable river estuaries where tidal barrages can be built
Nuclear	There is a plentiful supply of uranium-235 It is a concentrated energy resource: a large amount of energy can be generated from a relatively small amount of fuel Does not produce carbon dioxide or sulfur dioxide Reliable source of energy	Non-renewable Disposal of the nuclear waste produced in the reactor is problematic, since it has a long half-life and the emitted radiation is dangerous
Wind	Renewable Does not produce carbon dioxide or sulfur dioxide The area around the wind turbines can be used as farmland	Unreliable: cannot be used in very high winds and output is low on a calm day Wind turbines are noisy and some people consider them unsightly

Nuclear fusion

The process of nuclear fusion is carried out in the Sun. Hydrogen nuclei collide at great speed in the Sun and fuse together to form helium nuclei. This releases energy in the form of heat and light. The Sun is the source of all energy resources, except geothermal, nuclear and tidal.

Examination style question

1. The diagram below represents a hydroelectric system for generating electricity.

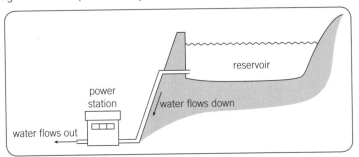

Answer the following questions, using words from this list.

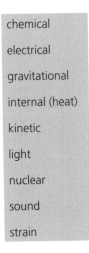

chemical

electrical

gravitational

internal (heat)

kinetic

light

nuclear

sound

strain

    a.   What sort of energy, possessed by the water in the reservoir, is the main source of energy for this system?

    b.   When the water flows down the pipe, it is moving. What sort of energy does it possess because of this movement?

    c.   The water makes the turbines in the power station rotate. What sort of energy do the turbines possess because of their rotation?

    d.   What sort of energy does the power station generate?

    e.   None of the energy transfer processes is perfect. In what form is most of the wasted energy released?

Cambridge IGCSE Physics 0625 Paper 2 Q4 June 2005

Work and power

The **work done** in **joules** by a force acting on an object = force × distance moved by the object in the direction of the force. The work done = energy transferred

work done (J) = force (N) × distance (m)

$$W \quad = \quad F \times d$$

The **power** in **watts** is the work done per second or the energy transformed per second.

$$\text{power (W)} = \frac{\text{energy (J)}}{\text{time (s)}}$$

$$P \quad = \quad \frac{E}{t}$$

Worked examples

1. A car engine produces a force of 2000 N while accelerating the car through a distance of 200 m in a time of 10 s.
   a. What is the work done on the car by the engine force?
   b. What is the power developed by the engine?

Answers

a. $W = F \times d$
   $= 2000 \times 200$
   $= 400\ 000$ J
   $= 400$ kJ

b. $P = \frac{E}{t}$
   $= \frac{400\ 000}{10}$
   $= 40\ 000$ W
   $= 40$ kW

Kinetic energy

Kinetic energy can be calculated from the formula:

$$KE = \frac{1}{2}mv^2$$

where m = mass in kg; v = velocity in m/s

When an object is lifted higher above the Earth's surface, work must be done.
Since work = force × distance
and force = weight of the object
work done = weight × height lifted
where weight = mass × gravitational field = mg

Gravitational potential energy

It follows that the change in **gravitational potential energy** (the work done in lifting the object) is given by the formula:

$$PE = mgh$$

where m = mass in kg; g = acceleration due to gravity 10 m/s²;
       h = change in height in m

Extended

Worked examples

1. A car of mass 1000 kg is travelling at a velocity of 20 m/s. Calculate its kinetic energy.

2. Calculate the change in potential energy of a 70 kg parachutist as she falls through a height of 100 m.

3. A ball of mass 0.5 kg is dropped from rest at a height of 5 m above the ground. Find its velocity when it hits the ground.

Answers

1. $KE = \frac{1}{2}mv^2$

$\quad = \frac{1}{2} \times 1000 \times 20^2$

$\quad = 200\ 000$ J

$\quad = 200$ kJ

2. $PE = mgh$

$\quad = 70 \times 10 \times 100$

$\quad = 70\ 000$ J

$\quad = 70$ kJ

3. $PE = mgh$

$\quad = 0.5 \times 10 \times 5$

$\quad = 25$ J

loss of PE = gain of KE

$v = \sqrt{(2KE/m)}$

$\quad = \sqrt{(2 \times 25/0.5)}$

$\quad = 10$ m/s

Examination style questions

1. An electric pump is used to raise water from a well, as shown below.

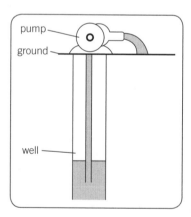

a. The pump does work in raising the water. State an equation that could be used to calculate the work done in raising the water.

b. The water is raised through a vertical distance of 8.0 m. The weight of water raised in 5.0 s is 100 N.

    i) Calculate the work done in raising the water in this time.

    ii) Calculate the power the pump uses to raise the water.

    iii) The energy transferred by the pump to the water is greater than your answer to (i). Suggest what the additional energy is used for.

Cambridge IGCSE Physics 0625 Paper 3 Q3 June 2006

Extended

2. A student wishes to work out how much power she uses to lift her body when climbing a flight of stairs.

Her body mass is 50 kg and the vertical height of the stairs is 4.0 m. She takes 20 s to walk up the stairs.

a. Calculate

　　i)   the work done in raising her body mass as she climbs the stairs,

　　ii) the output power she develops when raising her body mass.

b. At the top of the stairs she has gravitational potential energy. Describe the energy transformations taking place as she walks back down the stairs and stops at the bottom.

Adapted from Cambridge IGCSE Physics 0625 Paper 3 Q3 June 2007

1.8 Pressure

The **pressure** on a surface due to a force is the force on 1 m² of the surface.

$$\text{pressure (N/m}^2) = \frac{\text{force (N)}}{\text{area (m}^2)}$$

$$p = \frac{F}{A}$$

The unit of pressure, N/m² is also known as the **pascal (Pa)**.

Worked examples

1. A force of 10 kN acts on the surface of a liquid, of area 0.08 m². What is the pressure on the surface of the liquid?

2. A person of weight 600 N exerts a pressure of 40 kPa on the ground. What is the area of their feet?

3. The area of a dog's paw is 10 cm². The pressure under the paw is 50 kPa when it exerts half of its body weight on the paw. What is its weight?

Answers

1. $p = \frac{F}{A}$

$= \frac{10\ 000}{0.08}$

$= 125\ 000$ Pa

$= 125$ kPa

2. $A = \frac{F}{p}$

$= \frac{600}{40\ 000}$

$= 0.015$ m²

3. $F = p \times A$

$= 50\ 000 \times (10 \div 10\ 000)$

$= 50$ N

Total weight $= 2 \times 50 = 100$ N

Note: In question 1, 1 kN = 1000 N.

In question 3, area in cm² is converted to area in m² by dividing by (100 × 100) i.e. by 10 000.

Atmospheric (air) pressure can be measured with a **barometer**.

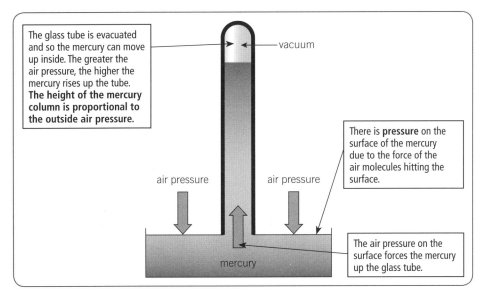

The glass tube is evacuated and so the mercury can move up inside. The greater the air pressure, the higher the mercury rises up the tube. **The height of the mercury column is proportional to the outside air pressure.**

vacuum

air pressure

air pressure

There is **pressure** on the surface of the mercury due to the force of the air molecules hitting the surface.

The air pressure on the surface forces the mercury up the glass tube.

mercury

Extended

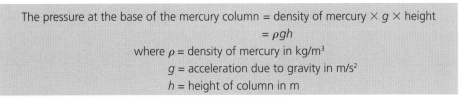

The pressure at the base of the mercury column = density of mercury × g × height
$$= \rho g h$$
where ρ = density of mercury in kg/m³
g = acceleration due to gravity in m/s²
h = height of column in m

The pressure below the surface of all liquids increases proportionally to depth. For example, the formula above can be used to calculate the pressure under the surface of the sea.

A **manometer** can be used to measure the pressure of a gas:

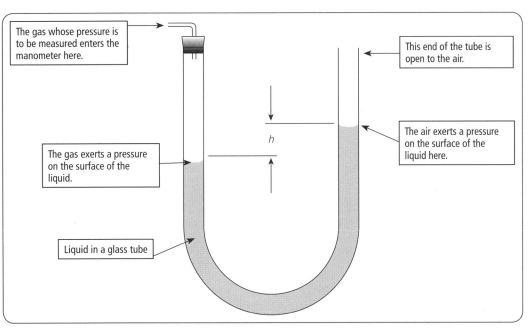

The gas whose pressure is to be measured enters the manometer here.

This end of the tube is open to the air.

The air exerts a pressure on the surface of the liquid here.

h

The gas exerts a pressure on the surface of the liquid.

Liquid in a glass tube

The pressure due to the gas = atmospheric pressure + $\rho g h$
Where ρ = density of liquid in kg/m³; h = difference in height of two liquid surfaces in m

Examination style questions

1. a. The diagram below shows two examples of footwear being worn by people of equal weight at a Winter Olympics competition.
Which footwear creates the greatest pressure below it, and why?

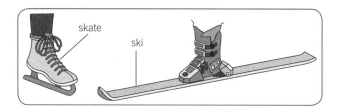

b. Drivers of high-sided vehicles, like the one below, are sometimes warned not to drive when it is very windy.

Suggest why they receive this warning.

Cambridge IGCSE Physics 0625 Paper 2 Q3 November 2005

2. a. A man squeezes a pin between his thumb and finger, as shown in the figure below.

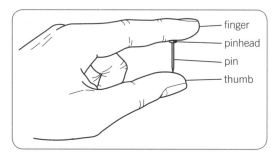

The finger exerts a force of 84 N on the pinhead.
The pinhead has an area of 6.0×10^{-5} m^2.

i) Calculate the pressure exerted by the finger on the pinhead.
ii) State the value of the force exerted by the pin on the thumb.
iii) Explain why the pin causes more pain in the man's thumb than in his finger.

b. The density of the water in a swimming pool is 1 000 kg/m^3. The pool is 3 m deep.

i) Calculate the pressure of the water at the bottom of the pool.
ii) Another pool has the same depth of water, but has twice the area.
State the pressure of the water at the bottom of this pool.

Cambridge IGCSE Physics 0625 Paper 31 Q6 June 2009

Summary questions on Unit 1

1. Copy and fill in the blanks.

The speed of an object can be calculated by working out _____

divided by _____.

The unit of speed is _____. Speed is equal to the _____ of a

distance–time graph. Velocity is equal to speed in a certain _____ and

it has the same _____. The acceleration of an object is equal to

the _____ _____ _____ of velocity and has units of _____.

Acceleration can be calculated from the _____ of a velocity–time graph. The

area under a velocity–time graph is equal to the _____.

2. Calculate the speed of an object that travels 10 m in 5 s.

3. The initial velocity of a car is 10 m/s. It reaches a velocity of 25 m/s in 5 s. What is its acceleration?

4. What is the gradient of a distance–time graph equal to?

5. What is the area under a velocity–time graph equal to?

6. What is the difference between mass and weight?

7. A pebble of mass 100 g is placed in a measuring cylinder containing 50 ml of water. The water level rises to 75 ml. What is the density of the pebble?

8. Copy and fill in the blanks.

A force can stretch an object, or change its _____. If small forces are applied to

a spring, the extension produced by a load is _____ _____ to the force

applied. This is true up to the limit of _____. A force can change the velocity of

an object by causing it to _____. The acceleration is _____ if the force

on the object is doubled and the mass stays constant. The net or overall force on an

object is called the _____ force. When forces act in opposite directions

they are _____ to find the resultant force. When forces act in the same

direction, they are _____ to find the resultant force. A force can change

the _____ in which an object is travelling without changing its _____.

This happens when the force is _____ to the direction in which the object is

travelling and it causes the object to move in a _____.

9. A force of 10 N produces an extension of 20 cm. What extension would be produced by a force of 2.5 N?

10. Copy and fill in the blanks.

The mass of an object is measured in _____. The weight of an object is

a type of _____ and is measured in _____. The weight can be

calculated by multiplying the _____ by the _____ _____

_____. The density of an object is equal to its mass divided by its

_____. The unit of density is _____ or _____.

11. Match the devices to the energy transformations.

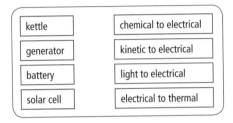

12. Calculate the kinetic energy of a ball of mass 0.2 kg thrown at a speed of 7 m/s.

13. Calculate the increase in gravitational potential energy of a boy of mass 50 kg climbing a flight of stairs of height 5 m.

14. Calculate the distance travelled by an object if the work done on it by a force of 20 N is 100 J.

15. What is the power of a kettle that transforms 100 J of electrical energy to internal energy in 0.5 s?

16. What is the pressure 3 m under the surface of the sea? Sea water has a density of 1200 kg/m³.

17. Calculate the pressure under a block of mass 0.3 kg when resting on an area of 5 cm².

Examination style questions on Unit 1

1. In a laboratory, an experiment is carried out to measure the acceleration of a trolley on a horizontal table, when pulled by a horizontal force.

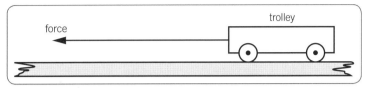

The measurements are repeated for a series of different forces, with the results shown in the table below.

force/N	4.0	6.0	10.0	14.0
acceleration $\frac{}{m/s^2}$	0.50	0.85	1.55	2.25

a. On the grid below, plot these points and draw the best straight line for your points.

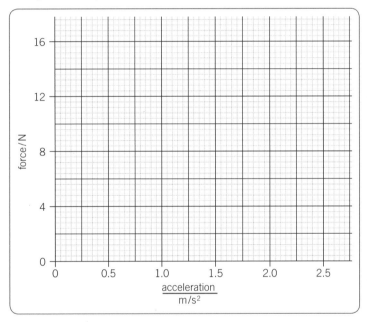

b. The graph shows that below a certain force there is no acceleration.
    i)   Find the value of this force.
    ii) A force smaller than that in **(b)(i)** is applied to the stationary trolley. Suggest what happens to the trolley, if anything.
c. Show that the gradient of your graph is about 5.7.
d. i)   State the equation that links resultant force F, mass m and acceleration a.
    ii) Use your gradient from **(c)** to find the mass of the trolley.
e. On the figure below, sketch a speed / time graph for a trolley with constant acceleration.

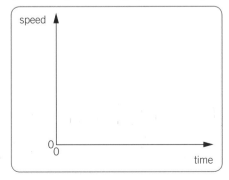

Cambridge IGCSE Physics 0625 Paper 31 Q1 June 2011

2. The period of the vertical oscillations of a mass hanging from a spring is known to be constant.
   a. A student times single oscillations with a stopwatch. In 10 separate measurements, the stopwatch readings were:

   1.8 s, 1.9 s, 1.7 s, 1.9 s, 1.8 s, 1.8 s, 1.9 s, 1.7 s, 1.8 s, 1.8 s.

   What is the best value obtainable from these readings for the time of one oscillation? Explain how you arrive at your answer.
   b. Describe how, using the same stopwatch, the student can find the period of oscillation more accurately.

   Cambridge IGCSE Physics 0625 Paper 31 Q1 June 2012

3. A bus travels from one bus stop to the next. The journey has three distinct parts. Stated in order they are:
   uniform acceleration from rest for 8.0 s,
   uniform speed for 12 s,
   non-uniform deceleration for 5.0 s.
   The graph below shows only the deceleration of the bus.

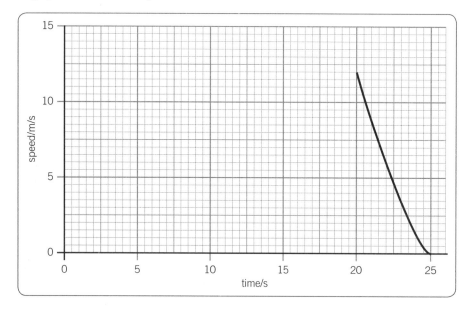

   a. Copy the graph and complete it to show the first two parts of the journey.
   b. Calculate the acceleration of the bus 4.0 s after leaving the first bus stop.
   c. Use the graph to estimate the distance the bus travels between 20 s and 25 s.
   d. On leaving the second bus stop, the uniform acceleration of the bus is 1.2 m/s².
      The mass of the bus and passengers is 4000 kg.
      Calculate the accelerating force that acts on the bus.
   e. The acceleration of the bus from the second bus stop is less than that from the first bus stop.
      Suggest two reasons for this.

   Cambridge IGCSE Physics 0625 Paper 3 Q1 June 2006

Extended

4. The diagram below shows a model car moving clockwise around a horizontal circular track.

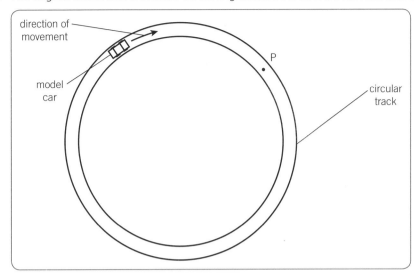

a.   A force acts on the car to keep it moving in a circle.
   i)   Copy the diagram and draw an arrow to show the direction of this force.
   ii)   The speed of the car increases. State what happens to the magnitude of this force.
b.   i)   The car travels too quickly and leaves the track at P. On your diagram, draw an arrow to show the direction of travel after it has left the track.
   ii)   In terms of the forces acting on the car, suggest why it left the track at P.
c.   The car, starting from rest, completes one lap of the track in 10 s. Its motion is shown graphically below.

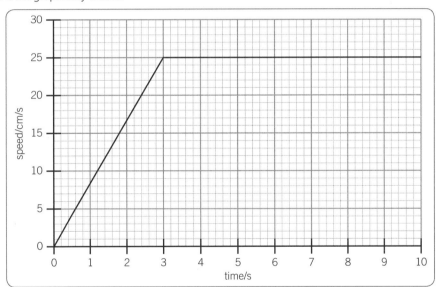

   i)   Describe the motion between 3.0 s and 10.0 s after the car has started.
   ii)   Use the diagram at the beginning of the question to calculate the circumference of the track.
   iii)   Calculate the increase in speed per second during the time 0 to 3.0 s.

Cambridge IGCSE Physics 0625 Paper 3 Q1 June 2007

5. a. Figure 1 shows a skier descending a hillside. Figure 2 shows the speed/time graph of his motion.

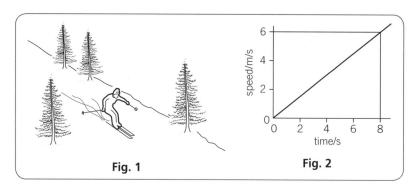

Fig. 1   Fig. 2

   i)   How can you tell that the acceleration of the skier is constant during the 8 s shown on the graph?

   ii) Calculate the acceleration of the skier.

   b. Another skier starts from rest at the top of the slope. As his speed increases, the friction force on the skier increases.

   i)   State the effect of this increasing friction force on the acceleration.

   ii) Eventually the speed of the skier becomes constant.

        What can be said about the friction force when the speed is constant?

   iii) 1. On the axes of Figure 3, sketch a possible speed/time graph for the motion of the second skier.

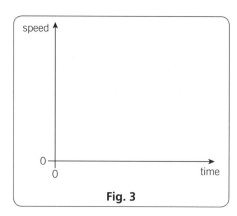

Fig. 3

        2. On your graph, mark with the letter A a region where the acceleration is not constant. Mark with the letter B the region where the speed is constant.

Cambridge IGCSE Physics 0625 Paper 31 Q3 June 2009

2 Thermal physics

2.1 Kinetic theory

KEY IDEAS

✓ The arrangement of particles varies in solids, liquids and gases
✓ The arrangement of particles affects the physical properties of a substance
✓ The kinetic energy of the particles is dependent on the temperature of the substance
✓ The particles in gases and liquids are in constant random motion
✓ When the particles of a gas collide with the walls of its container, they exert a pressure
✓ The pressure due to a gas at constant temperature is inversely proportional to its volume
✓ The highest energy particles escape the surface of a liquid due to evaporation, at temperatures below the boiling point

A molecule is one or more atoms (of the same or different elements, joined together by chemical bonds) capable of existing as a unit.

Molecular structure of solids, liquids and gases

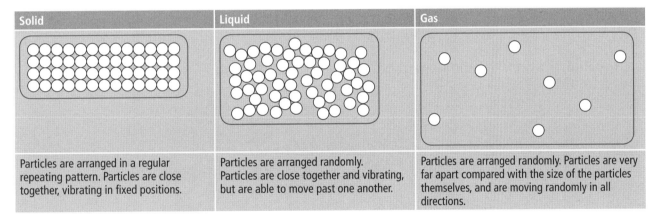

Solid	Liquid	Gas
Particles are arranged in a regular repeating pattern. Particles are close together, vibrating in fixed positions.	Particles are arranged randomly. Particles are close together and vibrating, but are able to move past one another.	Particles are arranged randomly. Particles are very far apart compared with the size of the particles themselves, and are moving randomly in all directions.

Physical properties of solids, liquids and gases

Solid	Liquid	Gas
Very difficult to compress	Difficult to compress	Easy to compress
Cannot flow	Can flow	Can flow
Keeps its shape	Takes the shape of the bottom of the container	Fills the entire container

Extended

Solids

In a solid, the particles are close together with strong bonds (attractive forces) between them. It is difficult to separate the particles and as a result solids are hard to break. As the particles are already tightly packed, they cannot be pushed closer together and so solids are **incompressible**.

Extended

Liquids

In a liquid, the particles are close together with weak bonds (attractive forces) between them. It is easy for the particles to move past one another and as a result, liquids can be poured and can flow. As the particles are already close together, they cannot be pushed much closer together and so liquids are **incompressible**.

Gases

In a gas, the particles are very far apart with negligible (practically zero) forces between them. It is easy for the particles to move past one another and the gas can flow. The particles can easily be pushed closer together and so gases are **compressible**.

Substances that can flow are called **fluids**. All liquids and gases are fluids.

Change in motion of molecules in a gas with temperature

As the temperature of a gas increases, the average kinetic energy of the particles increases and so the average speed of the particles increases.

Pressure due to a gas

The gas particles are moving fast in all directions. If the gas is placed in a sealed container, when the particles hit the sides of the container, they rebound. If they arrive with a momentum mv and leave with a momentum of $-mv$ (minus sign indicates direction), there is a momentum change of $2mv$. It follows that they exert a force on the walls of the container. Hence, there is a pressure on the side of the container equal to the force exerted divided by the area of the side. ($p = F/A$, see Section 1.8). If the temperature of the gas increases, the average speed of the particles and the number of collisions per second increases. Hence the force exerted on the container by the particles is greater and so the pressure increases.

Brownian motion: the movement of small particles in a gas or a liquid

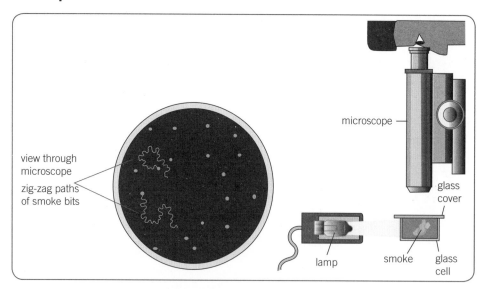

view through microscope

zig-zag paths of smoke bits

microscope

glass cover

lamp

smoke

glass cell

If smoke is introduced into a glass cell and observed through a microscope, the smoke particles can clearly be seen. The smoke particles appear to be in **random motion**. This is because they are constantly being hit by the other rapidly moving particles, in the air, in the glass cell.

This is evidence that the particles in air are in constant random motion. It shows that massive particles may be moved by light, fast-moving particles.

Extended

Boyle's Law: the variation of volume of a gas with applied pressure

Experimental arrangement

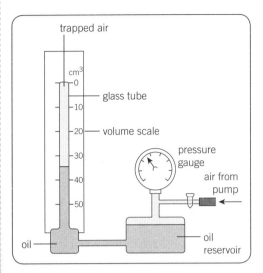

Experimental procedure

The pressure on a fixed mass of gas is varied by using a foot pump and measured using a pressure gauge at constant temperature. The volume of the gas is recorded as the pressure is increased.

Results

As the graph shows, when the pressure on the gas is increased, the volume of the gas decreases. If the pressure is doubled, the volume of the gas halves. The volume is **inversely proportional** to the pressure applied.

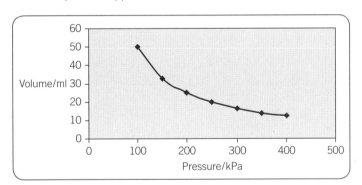

▲ A graph to show how the volume of a fixed mass of gas varies with pressure

Boyle's Law

For a fixed mass of gas at constant temperature, the volume is inversely proportional to the applied pressure.

$$pV = \text{constant}$$

Evaporation

Evaporation is the transformation of a liquid into a gas at a temperature **below the boiling point of the liquid.** The liquid particles have kinetic energy and are moving past one another, but some particles have more kinetic energy than the others. These faster particles may be moving fast enough to overcome the force of attraction between them and the other particles. The fast moving particles escape from the **surface** of the liquid and form a gas.

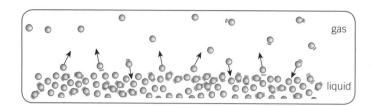

The rate of evaporation can be increased by:

*   increasing the temperature (e.g. put the liquid in a warmer room)
*   increasing the surface area (e.g. put the liquid in a wider container)
*   increasing the air flow over the surface of the liquid (e.g. blow on the surface).

Objects in contact with an evaporating liquid (e.g. a wet hand) cool as the liquid evaporates. Heat energy is transferred from the object to the evaporating liquid.

Examination style questions

1. The following diagram shows a plastic bottle containing only air. The bottle is sealed by the cap.

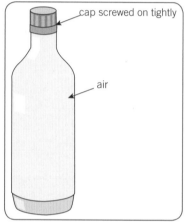

a.   Describe the motion of the molecules in the air inside the bottle.
b.   Describe and explain what happens to the pressure in the bottle as the temperature of the air increases.

2. An IGCSE student notices a puddle of water on the pavement as he leaves home on a warm day. When he returns one hour later, the puddle is only half as wide as before.
a.   State the name of the process causing the decrease in the size of the puddle.
b.   Explain how the process leads to the observed effect.

3. The diagram is a design for remotely operating an electrical switch using air pressure.

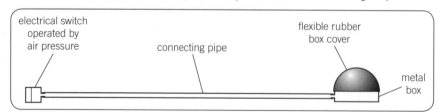

The metal box and the pipe contain air at normal atmospheric pressure and the switch is off. When the pressure in the metal box and pipe is raised to 1.5 times atmospheric pressure by pressing down on the flexible rubber box cover, the switch comes on.

a.   Explain in terms of pressure and volume how the switch is made to come on.
b.   Normal atmospheric pressure is 1.0×10^5 Pa. At this pressure, the volume of the box and pipe is 60 cm³.

Calculate the **reduction** in volume that must occur for the switch to be on.

c.   Explain, in terms of air particles, why the switch may operate, without the rubber cover being squashed, when there is a large rise in temperature.

Cambridge IGCSE Physics 0625 Paper 3 Q4 June 2008

4. a.   The diagram below shows the paths of a few molecules in air and a single dust particle. The actual molecules are too small to show on the diagram.

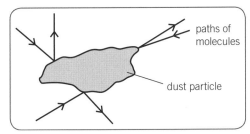

Explain why the dust particle undergoes small random movements.

b.   The diagram below shows the paths of a few molecules leaving the surface of a liquid. The liquid is below its boiling point.

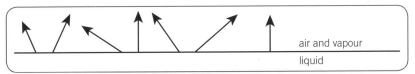

i)   State which liquid molecules are most likely to leave the surface.

ii)   Explain your answer to **(i)**.

Cambridge IGCSE Physics 0625 Paper 3 Q5 June 2005

5.   The diagram below shows a way of indicating the positions and direction of movement of some molecules in a gas at one instant.

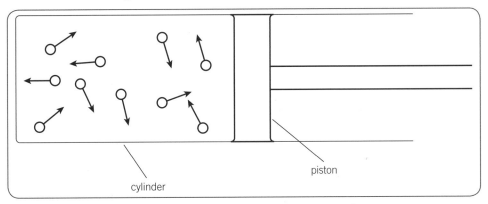

a.   i)   Describe the movement of the molecules.

ii)   Explain how the molecules exert a pressure on the container walls.

b.   When the gas in the cylinder is heated, it pushes the piston further out of the cylinder. State what happens to

i)   the average spacing of the molecules,

ii)   the average speed of the molecules.

c.   The gas shown in the diagram is changed into a liquid and then into a solid by cooling. Compare the gaseous and solid states in terms of

i)   the movement of the molecules,

ii)   the average separation of the molecules.

Cambridge IGCSE Physics 0625 Paper 3 Q5 November 2005

Extended

Practical question

An IGCSE student sets up the following experiment by wrapping the bulbs of three alcohol-in-glass thermometers with the same thickness of cotton wool. Thermometer 1 has dry cotton wool, thermometer 2 has cotton wool soaked in water and thermometer 3 has cotton wool soaked in acetone. She then investigates how the temperature of each thermometer varies with time.

▲ Thermometer 1 dry   ▲ Thermometer 2 water   ▲ Thermometer 3 acetone

The results of the experiment are shown in the table below:

Time/min	Temperature 1/°C	Temperature 2/°C	Temperature 3/°C
1	24	24	24
2	23	21	20
3	23	19	17
4	23	18	16
5	23	18	15

1. Compare and explain the trend in temperature for thermometer 1 and thermometer 2.

2. Compare and explain the results for thermometer 2 and thermometer 3.

2.2 Thermal properties

KEY IDEAS

✓ Most substances expand when heated, but water shows unusual behaviour between 0 °C and 4 °C
✓ The thermal expansion of metals can be used in bimetallic strips in applications such as thermostats, and must be accounted for in engineering applications
✓ The thermal expansion of liquids is used in liquid-in-glass thermometers
✓ The volume of a fixed mass of gas at constant pressure is dependent on temperature
✓ On the Celsius scale, thermometers are calibrated so they all read the same temperatures at 0 °C and 100 °C
✓ Specific heat capacity is a measure of a material's ability to store heat energy
✓ Latent heat of fusion is a measure of the heat energy required to melt a solid at its freezing point
✓ Latent heat of vaporisation is a measure of the heat energy required to boil a liquid at its boiling point

Thermal expansion of a solid

When solids are heated, they expand only a little, usually too little to see with the naked eye. However, in large structures such as bridges, buildings and railway tracks, this expansion could cause problems when there are large changes in temperature. To avoid this, railway tracks have small gaps between them to allow for expansion. If no gap was left, when the tracks expanded on a hot day they would be forced against each other and the track would bend and buckle.

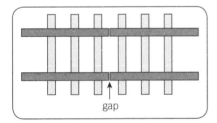

▲ Railway tracks on a cold day

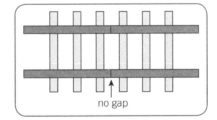

▲ Railway tracks on a very hot day

The expansion of metals can be used in a **bimetallic strip**, shown below.

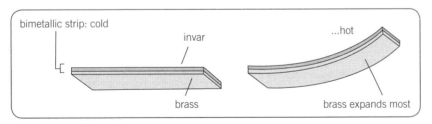

When the strip is heated, one metal expands more than the other and so the strip bends. The higher the temperature, the more the strip bends. In a **thermostat**, when the temperature rises, the bimetallic strip bends and the contacts separate, switching off the current to the heater. When the temperature falls, the strip goes back to its original position and the heater is switched on again. The temperature at which this occurs can be altered using the adjustable control knob.

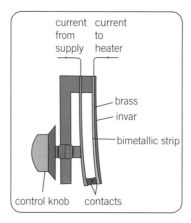

▲ Thermostat

Thermal expansion of a liquid

Liquids expand more than solids when heated, enough for the effect to be visible. This expansion is used in liquid-in-glass thermometers.

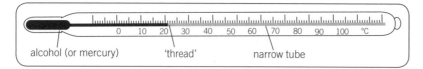

Extended

When the alcohol in the thermometer is heated by placing the bulb in a hot liquid, for example, the alcohol heats up and expands. This forces the alcohol up the narrow tube, so the thermometer gives a higher temperature reading. The narrow tube of constant cross-sectional area ensures that small changes in volume produce relatively large changes in length of the thread, increasing sensitivity.

Thermal expansion of a gas

Gases expand more than liquids when heated. When the temperature of a gas is increased, the particles gain kinetic energy and move around faster. This means they take up more space and so, if the gas is free to expand, it expands.

Extended

Relative expansion of solids, liquids and gases

The expansion of materials is dependent on the size of the forces between atoms. Solids have the greatest inter-atomic forces and therefore expand the least. Gases have the weakest forces and so they expand the most. Liquids expand more than solids, but less than gases.

Measurement of temperature

The **Celsius** scale of temperature is defined by its two fixed points: the freezing and boiling point of water. These are taken to be 0 °C and 100 °C (in standard atmospheric conditions).

The **Kelvin** scale of temperature is defined by the vibrations of particles. At **absolute zero**, (0 kelvin or **0 K**) particles have zero kinetic energy and no longer vibrate. Compared with the Celsius scale, this occurs at **–273 °C**. The two scales have the same sized degree, 1 °C is 274 K, so 2 °C is 275 K.

Kelvin temperature (K) = Celsius temperature (°C) + 273

Extended

Examples of thermometers

Liquid-in-glass	See expansion of liquids, on the previous page
Thermistor	A thermistor is an electrical component whose resistance decreases with temperature. It allows more current to flow when the temperature increases. The current reading is converted to a temperature reading on a digital meter.
Thermocouple	A thermocouple is made from two different types of metal, which are joined together to form two junctions. When there is a difference in temperature between the junctions, a small potential difference is created that depends on the temperature difference between the two junctions. The "hot junction" is placed in the substance whose temperature is to be measured. The potential difference between the hot and the cold junctions is then measured. The greater the temperature of the hot junction, the greater the potential difference. The potential difference is then converted to a temperature reading. Due to the small size of the junction, a thermocouple heats up quickly and so can be used to monitor rapidly changing temperatures. Due to the high melting point of the metal used to make the hot junction, a thermocouple can be used to measure high temperatures.

All thermometers calibrated on the Celsius scale agree at the fixed points of 0 °C and 100 °C, but not necessarily at temperatures in between these values. This is because the property of the material that is being used to measure temperature may not vary linearly. This affects the accuracy of the thermometer. The precision, or sensitivity, of the thermometer depends on how finely it is calibrated. For example, in an alcohol-in-glass thermometer, the divisions usually allow temperature to be measured to the nearest 1°C. Digital meters attached to a thermocouple or thermistor may give temperature to the nearest 0.1°C.

The range of a liquid-in-glass thermometer is limited by the freezing and boiling points of the liquid. A frozen thermometer will not work and if the liquid boils, it will explode! Thermistor and thermocouple thermometers may have greater ranges, but are not always as convenient as a liquid-in-glass thermometer.

Examination style questions

1. A manufacturer of liquid-in-glass thermometers changes the design in order to meet new requirements. Describe the changes that could be made to increase:
   a. the range of the thermometer
   b. the sensitivity of the thermometer

Adapted from Cambridge IGCSE Physics 0625 Paper 3 Q5b June 2006

2. The diagram below shows a fixed mass of gas trapped in a cylinder with a movable piston.

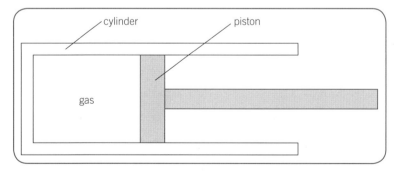

The temperature of the gas is increased.
   a. State what must happen to the piston, if anything, in order to keep the pressure of the gas constant.
   b. Explain your answer to part a.

Cambridge IGCSE Physics 0625 Paper 2 Q6b June 2007

3. a. State two changes that usually happen to the particles of a solid when the solid is heated.

   b. Most substances expand when they are heated.
      i) State one example where such expansion is useful.
      ii) State one example where such expansion is a nuisance, and has to be allowed for.

Cambridge IGCSE Physics 0625 Paper 2 Q5 June 2006

Thermal capacity

When heat energy is supplied to an object, its **internal energy** increases. This is because the molecules move faster so they have more kinetic energy. When an object's internal energy increases, the corresponding increase in temperature depends on its **thermal capacity**.

The thermal capacity of an object depends on the **material** from which it is made and its **mass**.

The **specific heat capacity (c)** of a material is the energy required to raise the temperature of 1 kg of the material by 1 °C.

For example, the specific heat capacity of water = 4200 J/(kg °C), which means that it takes 4200 J of energy to raise the temperature of 1 kg water by 1 °C.

Thermal capacity = mass × specific heat capacity

Thermal capacity = mc

Energy transferred = mass × specific heat capacity × temperature change

i.e.

$$E = mc\Delta T$$

Worked examples

1. Water has a specific heat capacity of 4200 J/(kg °C). How much heat energy must be supplied to raise the temperature of 1.5 kg of water by 10 °C?

2. 150 J of heat energy are required to raise the temperature of a 100 g block of metal by 5 °C. What is the specific heat capacity of the metal?

3. What is the increase in temperature of a 200 g block of metal of specific heat capacity 400 J/(kg °C) when 1500 J heat energy is supplied?

Answers

1. $E = mc\Delta T$
   $= 1.5 \times 4200 \times 10$
   $= 63\ 000$ J
   $= 63$ kJ

2. $c = \dfrac{E}{m\Delta T}$

   $= \dfrac{150}{0.100 \times 5}$

   $= 300$ J/(kg °C)

3. $\Delta T = \dfrac{E}{mc}$

$\quad = \dfrac{1500}{0.200 \times 400}$

$\quad = 19\,°C$

Note: In questions 2 and 3, the mass in g must be changed into kg by dividing by 1000 i.e.
$$100\,g = 0.100\,kg$$
$$200\,g = 0.200\,kg$$

An experiment to determine the specific heat capacity of water

Experimental arrangement

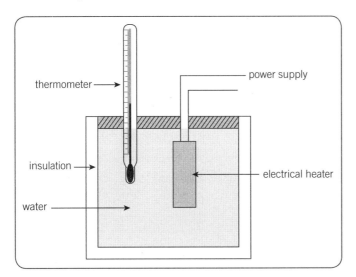

Procedure
Measure out 0.5 kg of water and pour into a beaker insulated with cotton wool. Place a thermometer in the water and cover it. Place an electrical heater of power 50 W in the water. Take the initial temperature of the water using the thermometer. Switch on the electrical heater and at the same time start a stop watch. Stir the water with the thermometer until the temperature has increased by 10 °C. Stop the stop watch and take the reading of the time taken for the water temperature to rise by 10 °C.

Processing the results
Time taken = 7 min = 420 s
Energy supplied by the heater = power × time = 50 × 420 = 21 000 J
Energy transferred to the water = $m \times c \times \Delta T$ = 0.5 × c × 10
Assuming all of the energy from the heater is transferred to the water,
0.5 × c × 10 = 21 000
$\qquad c = 4200\ J/(kg\,°C)$

It is unlikely that a student would be able to obtain such an accurate value for c in an experiment, as **some heat is always lost to the surroundings**.

Melting

The following graph shows how the temperature of a substance varies with time as heat energy is supplied at a constant rate.

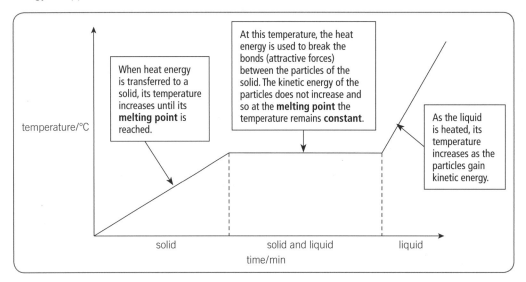

Solidification is the reverse of melting. A hot liquid loses heat to the surroundings which reduces its temperature. At the "melting point", the particles, instead of losing kinetic energy (causing a drop in temperature), arrange themselves into new "low energy" positions, i.e. the liquid becomes a solid.

The energy which must be put in to melt a solid at its melting point, or is given out when a liquid solidifies at its freezing point, is called the **latent heat of fusion**. There is no change in temperature.

Boiling

The following graph shows how the temperature of a substance varies as heat is supplied at a constant rate.

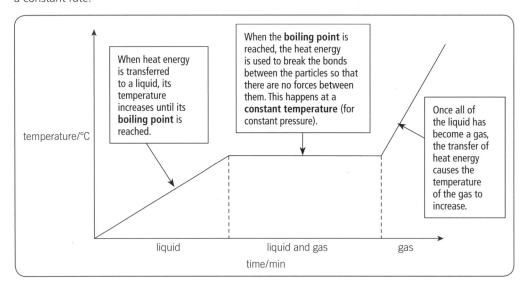

Boiling is different from **evaporation** in three ways:

- Boiling occurs at a fixed temperature, which depends on the substance being heated and its pressure.
- Evaporation can occur at all temperatures, including those below the boiling point.
- Evaporation decreases the temperature of the remaining liquid. The temperature of the liquid remains **constant** during boiling.

Condensation is the reverse of evaporation. If you breathe onto a cold window, you will see the water vapour (gas) in your breath turn back to liquid water as tiny droplets on the window. The window takes energy from the gas causing the particles to move to lower energy positions closer together, i.e. the gas becomes a liquid.

> The energy which must be put in to vaporise a liquid at its boiling point, or is given out when a gas at its boiling point condenses, is called the **latent heat of vaporisation**. There is no change in temperature.

The **specific latent heat of fusion** is the energy required to melt 1 kg of solid at its melting point, with no change in temperature.

The **specific latent heat of vaporisation** is the energy required to vaporise 1 kg of liquid at its boiling point, with no change in temperature.

For example, the specific latent heat of fusion for water = 300 000 J/kg, which means that it takes 300 000 J of energy to melt 1 kg of pure ice at 0 °C.

> Energy transferred = mass × specific latent heat
>
> i.e.
>
> $$E = \Delta mL$$

Worked examples

1. What mass of water is changed from liquid to gas when 30 kJ of energy is supplied (assuming that there is no change in temperature)? The specific latent heat of vaporisation of water is 2 300 000 J/kg.

2. How much energy is required to melt 0.3 kg of ice? The specific latent heat of fusion of ice is 330 000 J/kg.

Answers

1. $\Delta m = \dfrac{E}{L}$

$\qquad = \dfrac{30\ 000}{2\ 300\ 000}$

$\qquad = 0.013$ kg

2. $E = \Delta mL$

$\qquad = 0.3 \times 330\ 000$

$\qquad = 99\ 000$ J

$\qquad = 99$ kJ

Note: In question 1, kJ must be changed into J by multiplying by 1000.

Extended

An experiment to find the latent heat of fusion of ice

Experimental arrangement

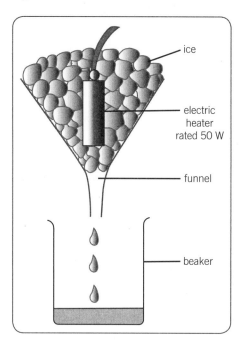

Procedure

Fill a funnel with ice and place a 50 W electric heater in the ice. Find the mass of the beaker using a balance. Place the beaker under the funnel and at the same time start a stop watch and switch on the electric heater. After 10 minutes, remove the beaker and switch off the heater. Find the mass of the beaker and water. Subtract the original mass of the beaker to find the mass of ice that has melted.

Processing the results

Energy supplied by the heater = power × time = $50 \times 600 = 30\,000$ J

Energy transferred to the water = $\Delta m \times L = 0.1 \times L$

Assuming only the energy from the heater is transferred to the melting ice,

$0.1 \times L = 30\,000$

$L = 300\,000$ J/kg

Inaccuracies in this experiment arise because some of the energy from the heater is lost to the surroundings and some energy from the surroundings is transferred to the ice. These two factors tend to cancel each other out and give a fairly accurate value for L.

Extended

An experiment to find the specific latent heat of vaporisation of water

Experimental arrangement

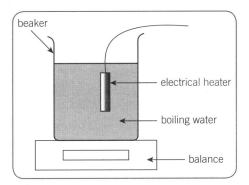

Procedure

Part fill a beaker with freshly boiled water and place on a balance. Place a 50 W electrical immersion heater in the water. Switch on the heater and wait for the water to boil. When the water is boiling, take the reading on the balance and at the same time start a stop watch. When the mass reading on the balance has decreased by 0.1 kg, take the reading on the stop watch.

Processing the results

Energy supplied by the heater = power × time = 50 × 4600
Energy transferred to the water = $\Delta m \times L = 0.1 \times L$
Assuming only the energy from the heater is transferred to the melting ice,
$$0.1 \times L = 230\,000$$
$$L = 2\,300\,000 \text{ J/kg}$$

Inaccuracies in this experiment arise because some of the energy from the heater is lost to the surroundings, which tends to give a value for L that is too large.

Examination style questions

1. An IGCSE student wishes to estimate the specific heat capacity of aluminium. She uses the experimental arrangement shown in the diagram.

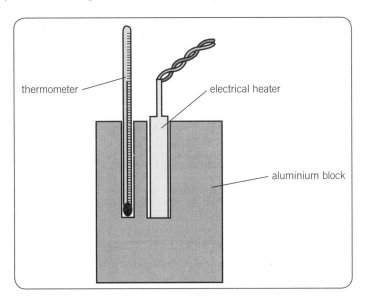

a. State the readings that she must take in order to calculate the specific heat capacity of aluminium.

Extended

b. Suggest whether the calculated value of specific heat capacity will be higher or lower than the actual value.
c. Explain your answer to part b.

Adapted from Cambridge IGCSE Physics 0625 Paper 3 Q4 June 2005

2. a. State two differences between evaporation and boiling.
b. Explain why energy is required to boil a liquid and why the temperature of the liquid remains constant during boiling.
c. A laboratory determination of the specific latent heat of vaporisation of water uses a 100 W heater to keep the water boiling at its boiling point. In 20 minutes, the mass of liquid water is reduced by 50 g. Calculate the value for the specific latent heat of vaporisation obtained from this experiment. Show your working.

Adapted from Cambridge IGCSE Physics 0625 Paper 3 Q4 June 2006

3. Some water is heated electrically in a glass beaker in an experiment to find the specific heat capacity of water. The temperature of the water is taken at regular intervals.

The temperature–time graph for this heating is shown below.

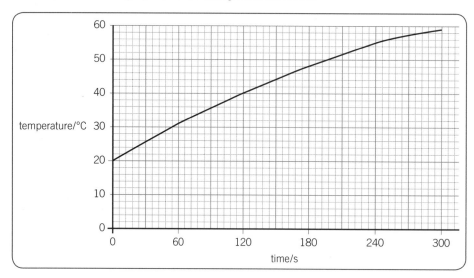

a. i) Use the graph to find
       1. the temperature rise in the first 120 s,
       2. the temperature rise in the second 120 s interval.
   ii) Explain why these values are different.

b. The experiment is repeated in an insulated beaker. This time, the temperature of the water increases from 20 °C to 60 °C in 210 s. The beaker contains 75 g of water. The power of the heater is 60 W. Calculate the specific heat capacity of water.
c. In order to measure the temperature during the heating, a thermocouple is used. Draw a labelled diagram of a thermocouple connected to measure temperature.

Cambridge IGCSE Physics 0625 Paper 3 Q4 November 2006

4. a. Suggest
       i) an example of a change of state resulting from the removal of thermal energy from a quantity of material,
       ii) the effect of this change of state on the temperature of the material.

b. Define the *thermal capacity* of a body.

Extended

c. A polystyrene cup holds 250 g of water at 20 °C. In order to cool the water to make a cold drink, small pieces of ice at 0 °C are added until the water reaches 0 °C and no unmelted ice is present.

[specific heat capacity of water = 4.2 J/(g °C), specific latent heat of fusion of ice = 330 J/g]

Assume no thermal energy is lost or gained by the cup.

i) Calculate the thermal energy lost by the water in cooling to 0 °C.
ii) State the thermal energy gained by the ice in melting.
iii) Calculate the mass of ice added.

Cambridge IGCSE Physics 0625 Paper 31 Q5 June 2012

Practical question

An IGCSE student is investigating the temperature rise of a beaker of water when heated by different methods. Beaker A is heated electrically and beaker B is heated by a Bunsen burner.

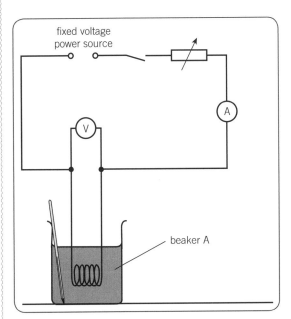
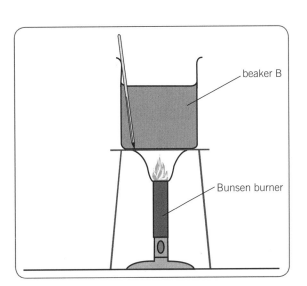

1. The student first records room temperature from the thermometer shown below.

Write down the value of room temperature.

2. The beakers are both heated for 1 minute. Beaker A reaches a temperature of 32 °C and beaker B reaches a temperature of 29 °C. Calculate the temperature rise in each beaker.

3. The student expected the temperature rise in beaker A and beaker B to be the same. Give two possible reasons why the results were different.

Adapted from Cambridge IGCSE Physics 0625 Paper 6 Q4 a & b June 2006

2.3 Transfer of thermal energy

KEY IDEAS

✓ Heat is transferred in solids by the vibrations of the molecules. This is conduction
✓ Heat is transferred in liquids and gases by the movement of the molecules. This is convection
✓ All objects emit and absorb heat through infrared radiation
✓ Black surfaces are the best emitters and the best absorbers of infrared radiation
✓ Silver surfaces are the worst emitters and the worst absorbers of infrared radiation

Conduction

An experiment to show that copper is a good conductor of heat

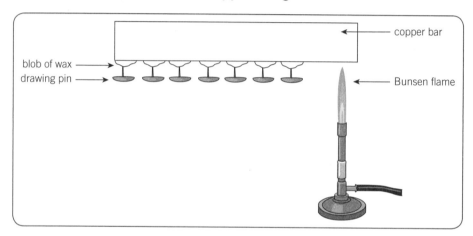

If a copper bar is heated at one end with a Bunsen flame, the drawing pins fall off one by one, beginning with the pin closest to the Bunsen flame. This is because as the metal **conducts** the heat from the hot end of the bar to the cold end, each of the blobs of wax melts in turn.

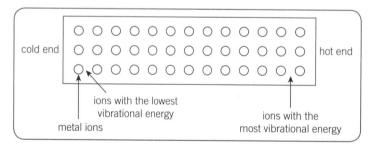

Extended

Metal atoms that have lost their free electrons are called **ions**. Sometimes metals are described as a lattice of metal ions in a "sea" of **free electrons**. Other solids are not usually good conductors of heat because they do not have free electrons to transfer heat energy from one molecule to another.

At the hot end of the metal bar, the ions gain energy and vibrate faster. The ions in a metal are close together and so these ions pass on their vibrations to neighbouring ions, which in turn start to vibrate faster. In this way, heat is transferred from the hot end to the cold end of the solid bar.

More significantly, metals also have free electrons in their structure which gain kinetic energy at the hot end of the bar. These free electrons pass on their kinetic energy through collisions with other electrons and metal atoms as they randomly diffuse through the metal. In this way, energy (heat) is conducted from the hot end of the bar to the cold end.

An experiment to show that water is a poor conductor of heat

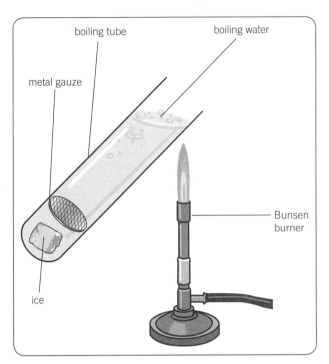

Ice is trapped at the bottom of the boiling tube with a piece of metal gauze. When the water at the top of the boiling tube is heated strongly, it boils. The ice at the bottom of the tube does not melt. This shows that water is a poor conductor of heat. However, if the ice is allowed to float normally, it melts quickly when the water is heated at the bottom of the test tube. This is because the water molecules can move, so the water heats by **convection**.

Water, like other liquids and non-metal solids, is a **poor conductor** of heat energy because its molecules do not have free electrons to easily pass on their kinetic energy to their neighbours, so the heat can only be transmitted through the vibration of the particles. Gases are **very poor conductors** of heat energy because their molecules are very far apart, so kinetic energy cannot be transmitted from one molecule to another.

Materials that are poor conductors are called **insulators**. For example, air is an insulator.

Convection

An experiment to demonstrate convection in water

A few crystals of potassium permanganate are placed at the bottom of a beaker of water. They dissolve and colour the water near them purple. When the water is heated, the purple water rises above the Bunsen flame, moves across and then falls at the other side of the beaker before returning to the flame to be heated again.

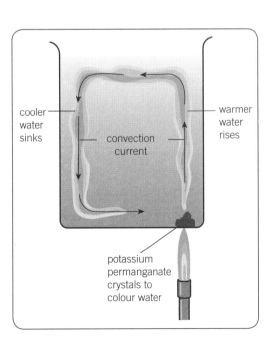

This movement of water is called a **convection current**.

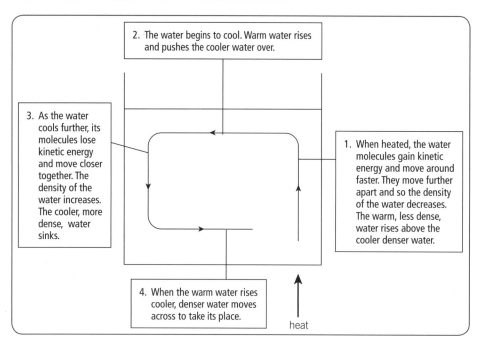

2. The water begins to cool. Warm water rises and pushes the cooler water over.

3. As the water cools further, its molecules lose kinetic energy and move closer together. The density of the water increases. The cooler, more dense, water sinks.

1. When heated, the water molecules gain kinetic energy and move around faster. They move further apart and so the density of the water decreases. The warm, less dense, water rises above the cooler denser water.

4. When the warm water rises cooler, denser water moves across to take its place.

heat

Convection can only take place in fluids (liquids and gases) where the particles are free to move. Materials that have trapped air in them such as cotton wool or bubble wrap are good insulators, because air does not conduct heat, and trapped air cannot convect heat either.

Radiation

All hot objects emit infrared (thermal) radiation, part of the **electromagnetic spectrum** of waves (see Section 3.3). Like all electromagnetic waves, infrared radiation can travel across a vacuum, which is why we are able to feel the heat of the Sun across the vacuum of space.

An experiment to demonstrate which type of surface is the best emitter of infrared radiation

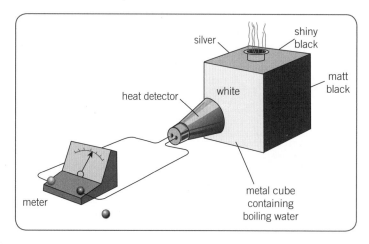

silver   shiny black

matt black

heat detector   white

meter

metal cube containing boiling water

Note that the sides are all at the same temperature. The greater the temperature, the higher the emission of radiation, which would affect the results.

Note that the sides of the cube are of equal area. The greater the area, the higher the emission of radiation, which would affect the results. The metal cube has its vertical sides painted with four different surfaces: matt black, shiny black, white and silver. It is filled with boiling water and a heat detector (a thermopile) placed at a constant distance from it. The cube is rotated so that each of the sides faces the detector in turn and the reading on the meter noted.

Extended

The readings on the meter vary as follows:

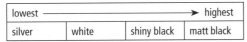

lowest			highest
silver	white	shiny black	matt black

Matt black surfaces are the best **emitters** of thermal radiation. Silver surfaces are the worst emitters of thermal radiation.

An experiment to demonstrate which colour surface is the best absorber of infrared radiation

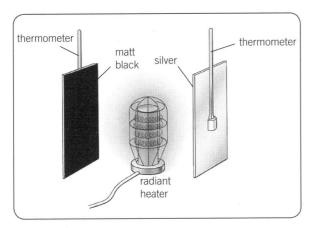

The initial readings are taken on the two thermometers. The radiant heater is switched on and the temperature on the two thermometers recorded at intervals of time.

Initially, the readings on the thermometers are the same, but after about 30 seconds the thermometer on the matt black surface gives a slightly higher reading than the thermometer on the silver surface. The temperature of the matt black surface continues to rise faster than the silver surface.

Matt black surfaces are the best **absorbers** of thermal radiation. Silver surfaces are the worst absorbers of heat radiation.

It follows that silver surfaces are the best reflectors of heat radiation, and matt black surfaces are the worst reflectors of heat radiation.

Examination style questions

1. The diagram below shows two containers, one painted matt black and the other shiny white. Both are filled with water, initially at the same temperature.

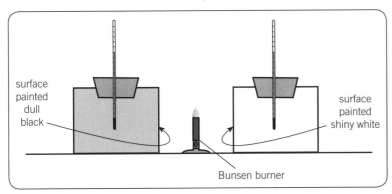

Describe how you would use the apparatus to determine which of the two surfaces is the better **absorber** of infrared radiation.

Adapted from Cambridge IGCSE Physics 0625 Paper 3 Q5a June 2007

2. Name the process by which thermal energy is transferred:
   a. from the Sun to the Earth.
   b. through the metal handle of a saucepan.

Adapted from Cambridge IGCSE Physics 0625 Paper 2 Q4a November 2006

Applications of conduction, convection and radiation

Vacuum flask

- The stopper is made of plastic, which is an insulator, to reduce heat flow by conduction. The stopper also stops heat flow by convection of air and stops heat flow by evaporation.
- The gap contains no air. So there are no particles to pass on the heat by conduction or convection.
- The silvered surfaces reflect infrared radiation and reduce heat flow by radiation.

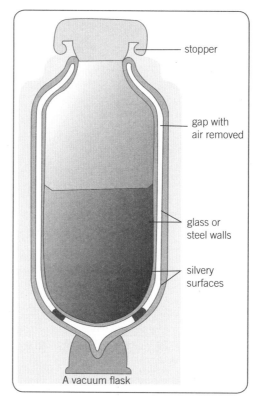

A vacuum flask

▲ A vacuum flask

Insulating the home

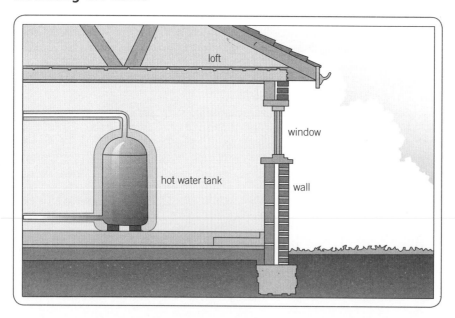

- "Lagging" around the hot water tank is made of plastic foam, which traps air to reduce heat loss by conduction (air is an insulator) and convection (the air is trapped and cannot move).
- Loft insulation is also made of fibres that trap air to reduce conduction and convection.

- An air cavity reduces heat exchange by conduction. Filling the air cavity with foam traps the air and also reduces convection.
- Double-glazed windows have air or another gas (argon or krypton) between the two layers of glass to reduce heat exchange by conduction and convection.

Sea breezes

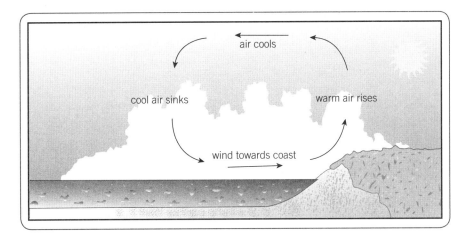

- During the day, **infrared radiation** from the Sun heats up the land more than the sea. The air above the land gets hotter than the air above the sea.
- The hot air above the land rises because it is less dense than the surrounding air.
- Cooler air from above the sea rushes in to take its place.
- This **convection current** causes a cool sea breeze to blow from the sea to the land.

Solar panel

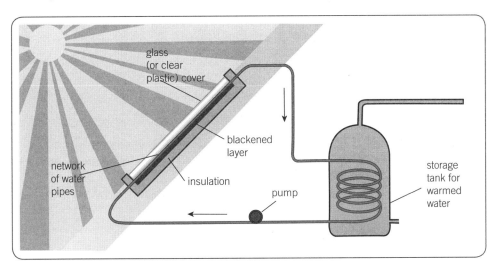

- The black surface of the solar panel absorbs **infrared radiation** from the Sun.
- The heat is then **conducted** through metal pipes to warm up water.
- The warm water rises to the top of the storage tank by **convection**.

Examination style questions

1. a. A crystal that dissolves slowly is put into some water in a beaker, as shown in the figure below.

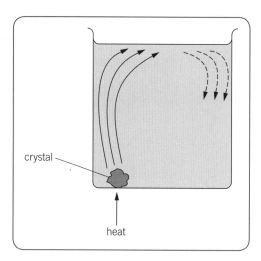

As it dissolves, the crystal colours the water around itself. When the beaker is heated, the coloured water moves as shown in the figure above.

i)   What name is given to this movement of the water?

ii) Describe why this movement happens.

b. A decorator, up a step-ladder painting the ceiling of a room, comments, "It is hotter up here by the ceiling than it is down on the floor."

Explain why his observation is correct.

Cambridge IGCSE Physics 0625 Paper 22 Q6 November 2012

2. An electric soldering iron is used to melt solder, for joining wires in an electric circuit. A soldering iron is shown below.

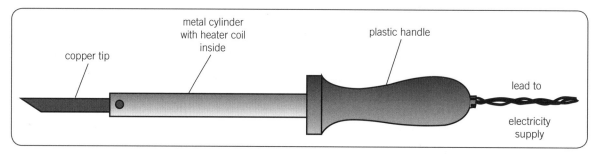

Solder is a metal which melts easily. The heater coil inside the metal cylinder heats the copper tip.

a. i)   Suggest why the tip is made of copper.
    ii) Suggest why the handle is made of plastic.

b. The heater coil is switched on. When the tip is put in contact with the solder, some of the heat is used to melt the solder.
    i)   State the process by which the heat is transferred from the copper tip to the solder.
    ii) By which process or processes is the rest of the heat transferred to the surroundings?

        conduction   convection   evaporation   radiation

c. A short time after switching on the soldering iron, it reaches a steady temperature, even though the heater coil is constantly generating heat.

    The soldering iron is rated at 40 W.

    What is the rate at which heat is being lost from the soldering iron?

        greater than 40 W
        equal to 40 W
        less than 40 W

Cambridge IGCSE Physics 0625 Paper 2 Q7 June 2007

3. a. The diagram below shows a copper rod AB being heated at one end.

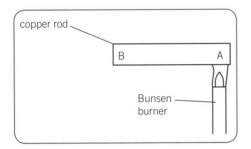

    i)   Name the process by which heat moves from A to B.
    ii) By reference to the behaviour of the particles of copper along AB, state how this process happens.

b. Give an account of an experiment that is designed to show which of four surfaces will absorb most heat radiation.

    The four surfaces are all the same metal, but one is a polished black surface, one is a polished silver surface, one is a dull black surface and the fourth one is painted white. Give your answer under the headings below:

    *   labelled diagram of the apparatus
    *   readings to be taken
    *   one precaution to try to achieve a fair comparison between the various surfaces.

Cambridge IGCSE Physics 0625 Paper 3 Q5 November 2006

Practical question

The IGCSE class carries out an experiment to investigate the effect of insulation on the rate of cooling of hot water. The apparatus is shown in the diagram below.

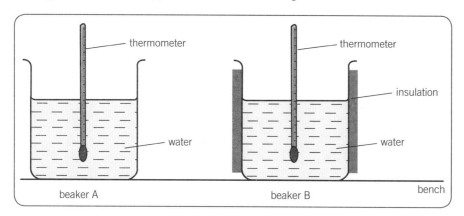

The students each have a stop watch, two glass beakers and some hot water. Beaker B is insulated.

A student fills beaker A about two-thirds full with hot water and takes the initial reading on the thermometer, θ. At the same time the student starts a stop clock. He takes readings of the temperature every 30 seconds for four minutes. He then repeats the experiment for beaker B. The results of the experiment are recorded below.

Beaker A	
$t/$	$\theta/$
0	85
30	71
60	62
90	55
120	52
150	49
180	47
210	45
240	44

Beaker B	
$t/$	$\theta/$
0	85
30	74
60	65
90	57
120	55
150	51
180	49
210	47
240	46

1. What unit is missing from the column headings? Plot graphs of temperature (θ) against time (t) for beaker A and B on the same axes. Label both axes and draw the curves of **best fit**. Start the temperature axis at 40 °C.

2. The experiment was designed to investigate the effect of insulation on the cooling of hot water. Suggest two ways to improve the experiment.

Adapted from Cambridge IGCSE Physics 0625 Paper 6 Q3 November 2005

Summary questions on Unit 2

1. Match each heat transfer mechanism to its description.

Conduction	An electromagnetic wave
Convection	Movement of hot fluid due to changes in density
Radiation	Most energetic particles escape from the surface of a liquid
Evaporation	Transfer of vibrational energy from particle to particle

2. Copy and fill in the blanks.

The specific latent heat of fusion is the energy required to _____ _____ kg of solid. The specific latent heat of vaporisation is the energy required to _____ _____ kg of liquid. The specific heat capacity is the energy required to increase the _____ of _____ kg of material by 1 °C.

3. Copy and complete the table.

	Solid	Liquid	Gas
Structure			
Arrangement of molecules		Close together, but able to move past one another	
Compressible?			Yes
Flows?		Yes	
Shape	Keeps its shape		

4. The diagram below shows the coastline at night. Copy the diagram and draw four arrows to show the direction of the convection current in the air. Label each arrow to explain why the air is moving in the direction you have indicated.

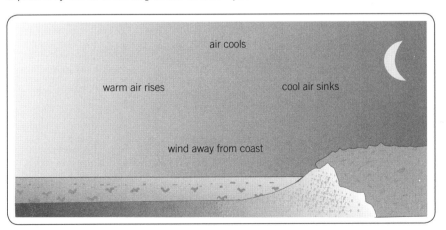

air cools

warm air rises   cool air sinks

wind away from coast

5. Write down the word equation used to calculate specific heat capacity.

Use the word equation to calculate the specific heat capacity of aluminium, if 4500 J of energy is required to raise the temperature of 500 g by 10 °C. Show your working. Give the unit of specific heat capacity.

6. Write down the word equation used to calculate specific latent heat of fusion.

Use the word equation to calculate the specific latent heat of fusion of ice, if 66 000 J of energy is required to melt 200 g of ice. Show your working. Give the unit of specific latent heat.

7. A sealed container with a moveable piston contains a fixed mass of dry air. The piston is used to reduce the volume of the gas without changing its temperature. Describe and explain the effect on the pressure of the gas.

The piston is now fixed so that it cannot move and the volume cannot change. The gas is heated. Describe and explain the effect on the pressure of the gas.

8. Crossword

Across:
1 Heat transfer by electromagnetic wave (9)
7 The energy required to increase the temperature of 1 kg of substance by 1 °C (8, 4, 8)
11 Scale of temperature (6)
12 Heat transfer by passing on vibrations from one molecule to another (10)
13 Metals are good conductors because of these in their structure (4, 9)
14 A substance in which there are no forces between molecules (3)
15 Type of thermometer (12)
19 A solid changes into a liquid at this constant temperature (7, 5)
21 The temperature at which all molecular vibrations stop (8, 4)

Down:
2 The relationship between the pressure and volume of a fixed mass of gas at constant temperature (9, 12)
3 This happens in a puddle of water on a warm day (11)
4 Liquid often used in liquid-in-glass thermometer (7)
5 Heat transfer through the movement of hot fluid (10)
6 Energy of movement (7)
8 A liquid changes into a gas at this constant temperature (7, 5)
9 The colour of surface that is the best absorber of thermal radiation (4, 5)
10 An instrument for measuring temperature (11)
16 The colour of surface that is the worst emitter of thermal radiation (6)
17 Stops all heat transfer apart from radiation (6)
18 The unit of energy (5)
20 Substance in which the molecules vibrate in fixed positions (5)

9. Draw a mind map (spider diagram), including all the important points from this unit. Use diagrams, colour coding and mnemonics to help you remember the key points. Ensure that you group the key ideas logically. When you have finished, ask someone to test you on the content of your mind map.

Examination style questions on Unit 2

1. The diagram below shows a student's attempt to estimate the specific latent heat of fusion of ice by adding ice at 0 °C to water at 20 °C. The water is stirred continuously as ice is slowly added until the temperature of the water is 0 °C and all the added ice has melted.

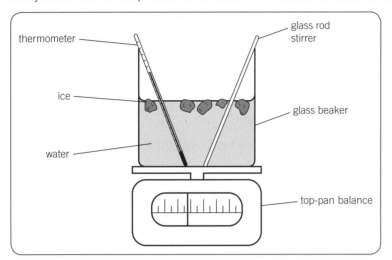

   a. Three mass readings are taken. A description of the first reading is given.
      Write down descriptions of the other two.
      Reading 1: the mass of the beaker + stirrer + thermometer
   b. Write down word equations which the student could use to find
      i)   the heat lost by the water as it cools from 20 °C to 0 °C,
      ii) the heat gained by the melting ice.
   c. The student calculates that the water loses 12 800 J and that the mass of ice melted is 30 g.
      Calculate a value for the specific latent heat of fusion of ice.
   d. Suggest two reasons why this value is only an approximate value.

Cambridge IGCSE Physics 0625 Paper 3 Q4 June 2004

2. During a period of hot weather, the atmospheric pressure on the pond in the diagram remains constant. Water evaporates from the pond, so that the depth h decreases.

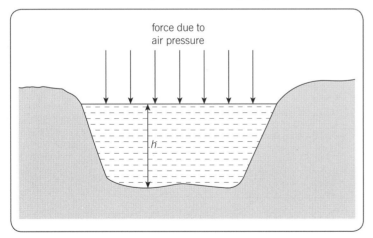

   a. Study the diagram and state, giving your reason, what happens during this hot period to
      i)   the force of the air on the surface of the pond,
      ii) the pressure at the bottom of the pond.

b. On a certain day, the pond is 12 m deep.
   i) Water has a density of 1000 kg / m³.
      Calculate the pressure at the bottom of the pond due to the water.
   ii) Atmospheric pressure on that day is 1.0×10^5 Pa.
      Calculate the total pressure at the bottom of the pond.
   iii) A bubble of gas is released from the mud at the bottom of the pond. Its initial volume is 0.5 cm³.
      Ignoring any temperature differences in the water, calculate the volume of the bubble as it reaches the surface.
   iv) In fact, the temperature of the water is greater at the top than at the bottom of the pond.
      Comment on the bubble volume you have calculated in **(b)(iii)**.

Cambridge IGCSE Physics 0625 Paper 31 Q3 June 2011

3. a. Explain why a liquid cools when evaporation takes place from its surface.
   b. The figure below shows five vessels each made of the same metal and containing water.
      Vessels A, B, C and D are identical in size and shape. Vessel E is shallower and wider. The temperature of the air surrounding each vessel is 20 °C.

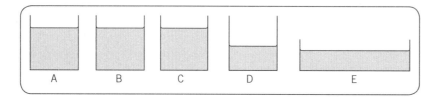

The table shows details about each vessel and their contents.

vessel	outer surface	volume of water/cm³	initial temperature of water/°C
A	dull	200	80
B	shiny	200	80
C	dull	200	95
D	dull	100	80
E	dull	200	80

The following questions are about the time taken for the temperature of the water in the vessels to fall by 10 °C from the initial temperature.
   i) Explain why the water in B takes longer to cool than the water in A.
   ii) Explain why the water in C cools more quickly than the water in A.
   iii) Explain why the water in D cools more quickly than the water in A.
   iv) Suggest **two** reasons why the water in E cools more quickly than the water in A.

Cambridge IGCSE Physics 0625 Paper 31 Q7 November 2012

3 Properties of waves

3.1 General wave properties

> ### KEY IDEAS
> ✓ Waves transfer energy without transferring matter
> ✓ Every wave can be described in terms of its frequency, wavelength, velocity and amplitude
> ✓ In a longitudinal wave the vibrations of the particles are parallel to the direction of motion. In a transverse wave, the vibrations are perpendicular to the direction of motion
> ✓ All the particles along a wavefront are at the same point in their vibration
> ✓ Water waves, produced by a ripple tank, can be reflected, refracted and diffracted
> ✓ Velocity = frequency × wavelength

A wave transfers energy from one place to another without transferring the particles of the medium. Individual particles vibrate (oscillate) about fixed positions.

There are two types of wave:

Longitudinal	Transverse
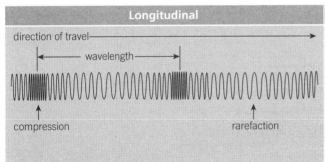	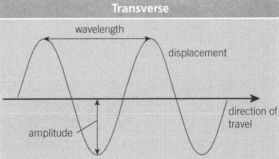
In a **longitudinal** wave the particles vibrate **parallel** to the direction of the wave. This leads to a series of **compressions** and **rarefactions**. In compressions, the particles are **closer together** than normal. In rarefactions, the particles are **further apart** than normal.	In a **transverse** wave the particles vibrate **perpendicular** to the direction of the wave. This leads to a series of **peaks** and **troughs**. At peaks, the particles are displaced **higher** than normal, at troughs they are displaced **lower** than normal.
An example of a longitudinal wave is sound in air.	An example of a transverse wave is light.

As shown on the two diagrams, the **wavelength is the distance between adjacent particles that are at the same point in their vibration** e.g. the distance between a compression and the next compression or the distance between a peak and the next peak.

The **amplitude** cannot easily be shown for the longitudinal wave, but for the transverse wave it is the **distance from the centre of a vibration to the peak**, measured in metres (m). The amplitude is the maximum displacement from the rest position.

The velocity of a wave is the distance travelled per second, measured in metres per second (m/s).

The frequency of a wave is the number of complete waves passing a point every second, measured in hertz (Hz).

Extended

Velocity = distance travelled by the wave in 1 second
Frequency = number of waves passing a point every second or the number of oscillations made by a particle on the wave every second.
The unit of frequency is **hertz (Hz)**

Velocity (m/s) = frequency (Hz) × wavelength (m)

$$v \quad = \quad f \quad \times \quad \lambda$$

Worked examples

1. A sound wave has a frequency of 10 000 Hz and a wavelength of 0.033 m. What is the speed of the wave?

2. A radio wave of speed 300 million m/s has a frequency of 600 MHz. What is its wavelength?

3. What is the frequency of an ultrasound wave of speed 1500 m/s and wavelength 0.05 m?

Answers

1. $v = f \times \lambda$
$= 10\ 000 \times 0.033$
$= 330$ m/s

2. $\lambda = \dfrac{v}{f}$

$= \dfrac{300\ 000\ 000}{600\ 000\ 000}$

$= 0.5$ m

3. $f = \dfrac{v}{\lambda}$

$= \dfrac{1500}{0.05}$

$= 30\ 000$ Hz
$= 30$ kHz

Note: 1 MHz = 1 million Hz = 1 000 000 Hz

 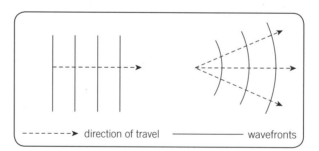

- - - - - - ▶ direction of travel ———————— wavefronts

Wavefronts can be represented as lines which are always **perpendicular** to the **direction of wave travel**. The distance between one wavefront and the next is one **wavelength**.

Ripple tank

The ripple tank has a vibrating bar attached to a motor, which can be used to set up waves of varying frequency in water.

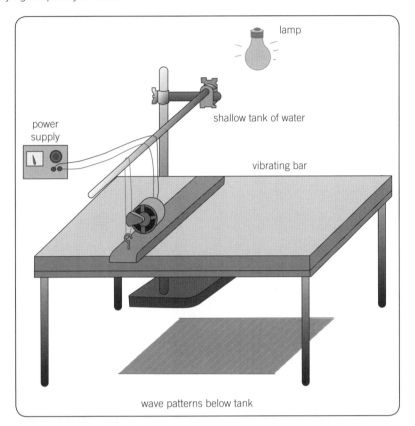

Water waves **reflect** at solid surfaces. There is no change in frequency, speed or wavelength on reflection.

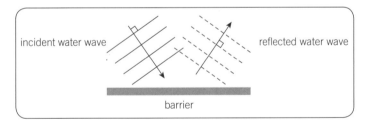

Wave theory suggests that each point on a wavefront can be considered to be a source of circular waves. These circular waves combine to make the wavefront and as they spread out, the wave travels forward.

Refraction

Water waves travel more slowly in shallow water.

As a wavefront AB approaches the boundary between the deep and shallow water, point A meets the shallower water first and this part of the wavefront **slows down**. Point B is still moving at the same speed as before. Wavefront AB turns clockwise which means that the new wavefront A'B' is now moving in a different direction. This change in direction is called **refraction**.

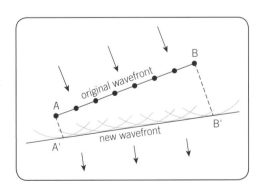

Diffraction

Water waves **spread out** as they go through a gap in a barrier. Only part of the wavefront can go through the gap. The waves that come from the original wavefront form a new curved wavefront. **Diffraction** through the gap occurs when the gap is greater than or equal to the wavelength of the waves.

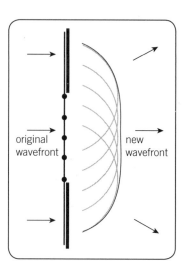

Extended

If the wavelength is approximately equal to the size of the gap, maximum diffraction occurs:

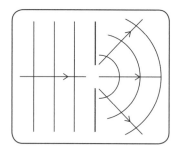

If the gap is much larger than the wavelength, little diffraction occurs:

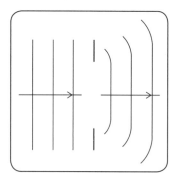

Similarly, with diffraction at an edge, the closer the wavelength is in size to the diffracting object, the greater the diffraction.

Examination style questions

1. The diagram shows the waveform of the note from a bell. A grid is given to help you take measurements.

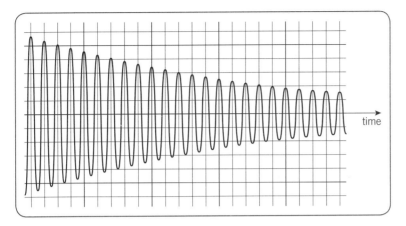

a. i)   State what, if anything, is happening to the loudness of the note.
    ii) State how you deduced your answer to **(a)(i)**.
b. i)   State what, if anything, is happening to the frequency of the note.
    ii) State how you deduced your answer to **(b)(i)**.
c. i)   How many oscillations does it take for the amplitude of the wave to decrease to half its initial value?
    ii) The wave has a frequency of 300 Hz.

    1. What is meant by *a frequency of 300 Hz*?

    2. How long does 1 cycle of the wave take?

    3. How long does it take for the amplitude to decrease to half its initial value?

d. A student says that the sound waves, which travelled through the air from the bell, were longitudinal waves, and that the air molecules moved repeatedly closer together and then further apart.
    i)   Is the student correct in saying that the sound waves are longitudinal?
    ii) Is the student correct about the movement of the air molecules?
    iii) The student gives light as another example of longitudinal waves. Is this correct?

Cambridge IGCSE Physics 0625 Paper 2 Q6 June 2009

2.
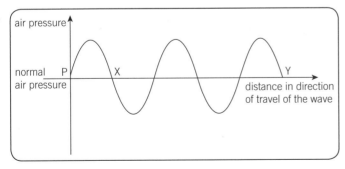

The diagram above shows how the pressure in air varies at one instant with distance along a sound wave.
a. Mark with a letter **C** on the diagram a point where there is a compression. Mark with a letter **R** on the diagram a point where there is a rarefaction.
b. Describe the motion of a group of air particles along the wave shown in the diagram.
c. The sound wave has a speed of 330 m/s and a frequency of 500 Hz. Calculate the wavelength of the wave and find the distance PX.

Adapted from Cambridge IGCSE Physics 0625 Paper3 Q7b-d June 2006

KEY IDEAS

✓ Virtual images cannot be focused on a screen
✓ When light is reflected, the angle of incidence (i) is equal to the angle of reflection (r)
✓ When light slows down it bends towards the normal. $\frac{\sin i}{\sin r} = n$ where

$$n = \frac{\text{speed of light in vacuum}}{\text{speed of light in medium}}$$

✓ $n = \frac{1}{\sin c}$, where c = critical angle
✓ Convex converging lenses produce different images, depending on the distance of the object from the lens
✓ When white light is refracted by a triangular prism it splits into the colours of the spectrum. This is called dispersion
✓ The electromagnetic spectrum is the name given to all the waves that travel at the speed of light in a vacuum. Their properties vary with frequency

Reflection of light

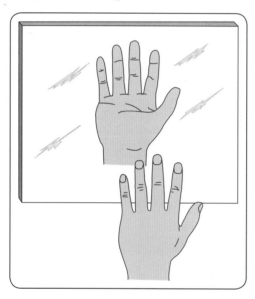

- The image is as far behind the mirror as the object is in front.
- The image is the same size as the object.
- The image is laterally inverted, which means left and right are swapped around.

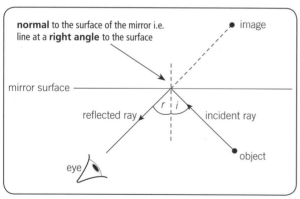

- angle of reflection = r
- angle of incidence = i
- The image can only be seen by the eye. It cannot be focused onto a screen, which is why it is **virtual**.
- The reflected ray appears to come from a point behind the mirror. This is the location of the **virtual image**.

Extended

▲ Ray diagram

- The angle of incidence is equal to the angle of reflection
- The incident ray, reflected ray and normal lie in the same plane

Refraction of light

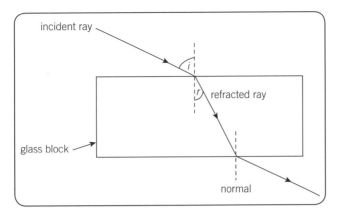

- i = angle of incidence
- r = angle of refraction
- As the ray of light enters the glass block it **slows** down and **bends towards the normal** so $i > r$.
- As the ray of light leaves the glass block it speeds up and bends **away from the normal**.
- The frequency of light is unchanged as it travels from one medium to another. As the light speeds up the wavelength increases. As the light slows down the wavelength decreases.

When light travels from an optically **less dense medium to a more dense** medium, e.g. from air to glass, it **slows down** and so **bends towards the normal**. When light travels from a more dense to a less dense medium it **speeds up** and **bends away from the normal**.

If the angle of incidence is 90°, the ray enters along the normal to the surface of the glass block. The light slows down, but does **not** change direction. As it leaves, it speeds up, but does not change direction.

The **refractive index** of a material (medium) is related to how dense it is. Generally, the denser the material, the higher the refractive index.

Refractive index of a medium, $n = \dfrac{\text{speed of light in a vacuum}}{\text{speed of light in the medium}}$

n is a constant for the material and has no unit.

The relationship between the angle of incidence i and the angle of refraction r is given by:

$$n = \frac{\sin i}{\sin r}$$

Worked examples

1. A ray of light is incident in air on a glass block at an angle of 30° to the normal. The refractive index of glass is 1.3. Calculate the angle of refraction.

2. A ray of light in air incident is on a transparent plastic block at an angle of 45° to the normal. The angle of refraction is 30°. Find the refractive index of the plastic.

3. The angle of refraction for a ray of light entering a diamond is 15°. What is the angle of incidence if the refractive index is 2.8?

Extended

Extended

Answers

1. $\sin r = \dfrac{\sin i}{n}$

$\sin r = \dfrac{\sin 30°}{1.3}$

$r = 22.6°$

2. $n = \dfrac{\sin i}{\sin r}$

$= \dfrac{\sin 45°}{\sin 30°}$

$= 1.4$

3. $\sin i = n \times \sin r$

$= 2.8 \sin 15°$

$i = 46.4°$

Critical angle

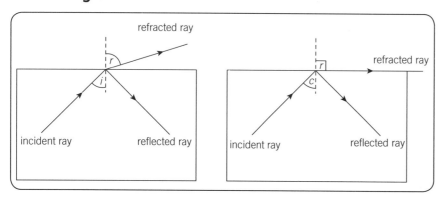

As the angle of incidence increases, the angle of refraction also increases, until $r = 90°$. When $r = 90°$, the **angle of incidence = the critical angle**. If the angle of incidence is increased beyond the critical angle, **total internal reflection** occurs.

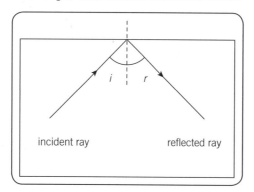

▲ The law of reflection $i = r$ is obeyed when light is totally internally reflected. The angle of incidence > critical angle.

Reflection may occur every time a ray of light is incident on a boundary between two media, but **total** internal reflection can **only** occur when the incident angle is **greater than the critical angle**.

Calculating refractive index from the critical angle

$$n = \frac{1}{\sin c}, \qquad \text{where} \quad n = \text{refractive index}$$
$$c = \text{critical angle}$$

Worked examples

1. What is the refractive index of a material whose critical angle is 40°?

2. Calculate the critical angle for a material of refractive index 1.3.

Answers

1. $n = \dfrac{1}{\sin c} = \dfrac{1}{\sin 40°} = 1.56$

2. $\sin c = \dfrac{1}{n} = \dfrac{1}{1.3}$

$c = \sin^{-1}\left(\dfrac{1}{1.3}\right) = 50°$

Total internal reflection can be used in **optical fibres**. An optical fibre has a thin glass cylindrical core coated with a transparent material of lower refractive index (cladding).

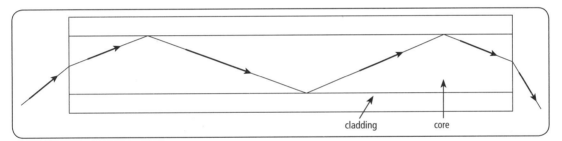

cladding   core

The **cladding** has a **lower refractive index** than the **core**, i.e. it is less dense. This means that **total internal reflection** will occur for all rays of light that hit the boundary between the core and cladding at an angle greater than the **critical angle**.

Optical fibres can be used to communicate signals, for example telephone conversations, or in medical applications such as endoscopy (a method of examining inside the body).

Examination style questions

1. The diagram (right) shows a ray of light OPQ passing through a semi-circular glass block.
   a. Explain why there is no change in the direction of the ray at P.
   b. State the changes, if any, that occur to the speed, wavelength and frequency of the light as it enters the glass block.
   c. At Q some of the light in ray OPQ is reflected and some is refracted. On a copy of the diagram, draw in the approximate positions of the reflected ray and the refracted ray. Label these rays.
   d. The refractive index for light passing from glass to air is 0.67. Calculate the angle of refraction of the ray that is refracted at Q into air.

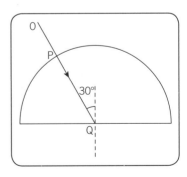

Cambridge IGCSE Physics 0625 Paper 3 Q6 June 2005

2. a. A ray of light in air travels across a flat boundary into glass. The angle of incidence is 51°. The angle of refraction is 29°.

    i) Draw a labelled diagram to illustrate this information.

    ii) Calculate the refractive index of the glass.

  b. A ray of light in glass travels towards a flat boundary with air. The angle of incidence is 51°. This ray does not emerge into the air.

  State and explain what happens to this ray.

Cambridge IGCSE Physics 0625 Paper 31 Q8 November 2012

3. The diagram below shows a ray of light, from the top of an object PQ, passing through two glass prisms.

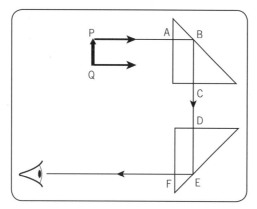

a. Copy and complete the path through the two prisms of the ray shown leaving Q.

b. A person looking into the lower prism, at the position indicated by the eye symbol, sees an image of PQ. State the properties of this image.

c. Explain why there is no change in direction of the ray from P at points A, C, D and F.

d. The speed of light as it travels from P to A is 3×10^8 m/s and the refractive index of the prism glass is 1.5. Calculate the speed of light in the prism.

e. Explain why the ray AB reflects through 90° at B and does not pass out of the prism at B.

Cambridge IGCSE Physics 0625 Paper 3 Q6 November 2006

Practical question

A student investigates the refraction of light through a transparent block. He places the transparent block on a sheet of plain paper, largest face down, and draws a line round the block. He draws a line to represent an incident ray and places two pins **W** and **X** in the line. The diagram below shows the outline of the block and the incident ray.

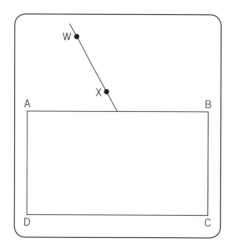

1. On a copy of the diagram, draw a normal to line **AB** at the point where the incident ray meets the block. The incident ray is drawn on the diagram. The positions of the two pins **W** and **X** that mark the incident ray are shown.

2. Measure the angle of incidence *i*.

3. Draw in the refracted ray with an angle of refraction of 20°. Continue this line until it meets the line **CD**.

4. The ray emerges from the block in a direction that is parallel to the incident ray. Draw in this emergent ray.

5. Two pins **Y** and **Z** are placed so that the pins **W** and **X**, viewed through the block, and the pins **Y** and **Z** all appear exactly in line with each other. Mark on your diagram, with the letters **Y** and **Z**, where you would place these two pins.

Cambridge IGCSE Physics 0625 Paper 6 Q5 June 2005

Dispersion of light

Light of a single frequency is said to be monochromatic. White light is made up of a range of different frequencies. We see different frequencies of light as different colours and so we can say that white light is made up of many different colours. This can be demonstrated experimentally using a triangular glass prism.

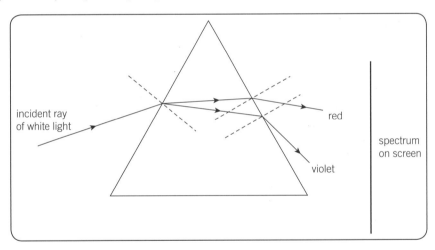

When white light passes through a triangular glass prism, it **refracts** at each surface and is deviated through a large angle. Of all the colours that make up white light, **violet** travels **slowest** in glass and **red** travels **fastest** in glass.

This means that violet is bent through the largest angle and red through the smallest. The other colours of visible light appear in between and so a **spectrum** is seen on the screen.

The refractive index of glass is higher for violet light than for red light.

When light spreads out into the colours of the spectrum, this is called **dispersion**.

Converging lens

When parallel rays of light from a distant source pass through a **convex converging** lens, they are focused to a point, which is called the **principal focus**. The principal focus lies on the principal axis, which is an imaginary line perpendicular to the plane of the lens. The distance between the principal focus and the centre of the lens is called the **focal length**.

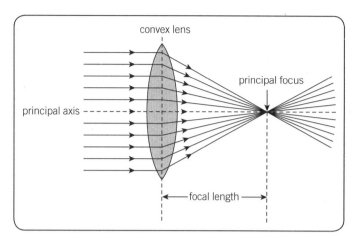

How to draw a ray diagram

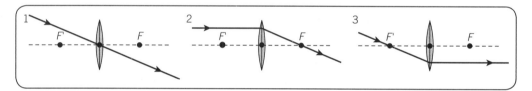

As these three diagrams show, each ray diagram should include **three** rays. Where the three rays meet, an image is formed. Each ray must have an **arrow** on it to indicate the **direction** of travel of the light.

Object between *F* and 2*F*

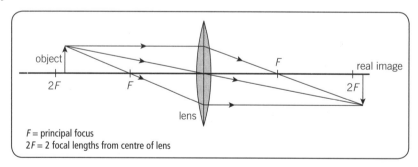

F = principal focus
2*F* = 2 focal lengths from centre of lens

When the object is between *F* and 2*F*, the image is **real** (can be focused on a screen), **inverted** (upside down) and **magnified** (bigger than the object).

This arrangement of a convex lens is used in a **projector**.

Object beyond 2F

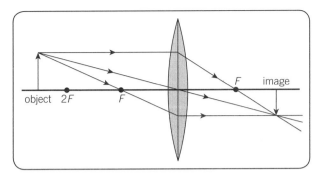

When the object is beyond 2F, the image is **real**, **inverted** and **diminished** (smaller than the object).

This arrangement of a convex lens is used in a **camera**.

Object between the centre of the lens and F

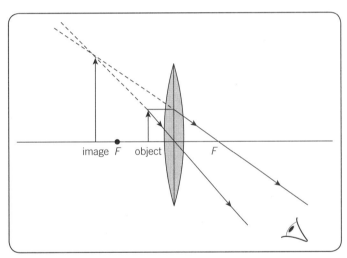

When the object is between the centre of the lens and F, the image is **virtual** (cannot be focused on a screen, but can be seen by looking into the lens), **magnified** and **upright** (the same way up as the object).

This arrangement of a convex lens is used in a **magnifying glass**.

Examination style questions

1. a. Copy the diagram below that shows a ray of blue light shining onto a glass prism.

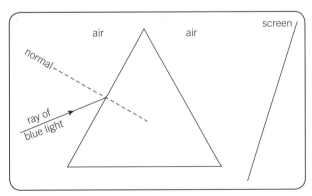

With the aid of a straight edge, draw a possible path of the ray through the prism and into the air until it reaches the screen.

Extended

b. When a ray of white light passes through the prism, it spreads into a spectrum of colours that can be seen on the screen.
   i)   Which of the following is the name of this spreading effect?
        convergence, diffraction, dispersion, reflection
   ii) Which colour is deviated **least** by the prism?
   iii) Which colour is deviated **most** by the prism?

Cambridge IGCSE Physics 0625 Paper 2 Q7 June 2006

2. Copy the following diagram. It is drawn to scale and shows an object PQ and a convex lens.

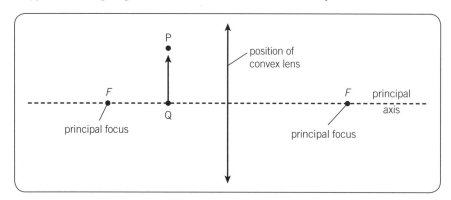

a. Draw two rays from the top of the object P that pass through the lens. Use these rays to locate the top of the image. Label this point **T**.
b. Draw an eye symbol to show the position from which the image T should be viewed.

Cambridge IGCSE Physics 0625 Paper 3 Q7 November 2005

3. The diagram shows an object, the tip of which is labelled O, placed near a lens L.

   The two principal foci of the lens are F_1 and F_2.

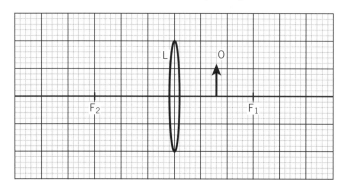

a. On the diagram above, draw the paths of two rays from the tip of the object so that they pass through the lens and continue beyond.

   Complete the diagram to locate the image of the tip of the object. Draw in the whole image and label it I.

b. Describe image I.

Cambridge IGCSE Physics 0625 Paper 3 Q6 June 2008

3.3 The electromagnetic spectrum

All electromagnetic waves can travel through a vacuum. They all travel at the same speed in a vacuum: 3×10^8 m/s, but travel more slowly through other materials (in air, electromagnetic waves travel at approximately the same speed as in a vacuum). Each section of the electromagnetic spectrum has its own uses and features.

Region	Frequency range/Hz	Wavelength range/m	Applications
Radio	$10^5 - 10^{10}$	$10^4 - 10^{-2}$	Electromagnetic oscillations produced by electrical circuits. Received by aerial. Used for communication.
Microwave	$10^{10} - 10^{11}$	$10^{-2} - 10^{-3}$	Used for rapid heating in microwave oven, sending signals between satellites and the Earth (e.g. mobile phones), radar imaging.
Infrared	$10^{11} - 10^{14}$	$10^{-3} - 10^{-6}$	All hot objects produce infrared radiation. Used in burglar alarms, night vision goggles and optical fibre communication.
Visible	10^{14}	10^{-7}	Produced by very hot objects such as the Sun. Detected by the eye and photographic film. Used in optical fibre communication.
Ultra-violet	$10^{15} - 10^{17}$	$10^{-7} - 10^{-9}$	Causes fluorescence in some materials (absorption of UV and emission of visible light). Used with sun beds to produce a sun tan and in identifying security marked valuables.
X-radiation	$10^{17} - 10^{19}$	$10^{-9} - 10^{-11}$	Blackens photographic film. Used to take images of bones in medical diagnosis (X-rays penetrate flesh more so than bone). Dangerous in high doses since ionising (see Section 5.2).
Gamma	$10^{19} - 10^{22}$	$10^{-11} - 10^{-14}$	Produced in the nuclei of radioactive elements. Used in medical diagnosis and therapy. Dangerous in high doses since ionising (see Section 5.2).

increasing energy

Examination style question

1. The diagram below shows the parts of the electromagnetic spectrum.

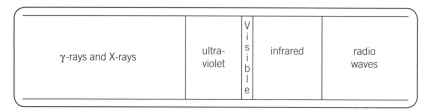

| γ-rays and X-rays | ultra-violet | V i s i b l e | infrared | radio waves |

a. Name one type of radiation that has
   i)   a higher frequency than ultra-violet
   ii) a longer wavelength than visible light.
b. Some γ-rays emitted from a radioactive source have a speed in air of 3.0×10^8 m/s and a wavelength of 1.0×10^{-14} m. Calculate the frequency of the γ-rays.
c. State the approximate speed of infrared waves in air.

Cambridge IGCSE Physics 0625 Paper 3 Q7 June 2005

3.4 Sound

Vibrating objects such as this tuning fork vibrate the molecules in the air and form a series of **compressions** (higher pressure than normal where the air molecules are close together) and **rarefactions** (lower pressure than normal where the air molecules are far apart).

The faster the tuning fork vibrates, the higher the **frequency** of the sound wave and hence the higher the **pitch.** If the tuning fork vibrates with greater **amplitude**, the sound is **louder**.

Sound waves in air are **longitudinal**. We hear sound when sound waves cause our ear drums to vibrate. We can hear a range of frequencies from about 20 Hz to 20 kHz. As we get older, this upper limit of 20 kHz gets lower. Ultrasound is transmitted by waves above the frequency range that humans can hear.

▲ Tuning fork

Sound waves cannot travel through a vacuum. They must be transmitted through the vibrations of particles in a medium. The closer together the particles in the medium, the faster the sound waves travel. So sound travels faster in solids than in the air.

For example:

air	water	metal
slowest ————————————————▶		fastest
330 m/s	1500 m/s	5000 m/s

An experiment to find the speed of sound in air

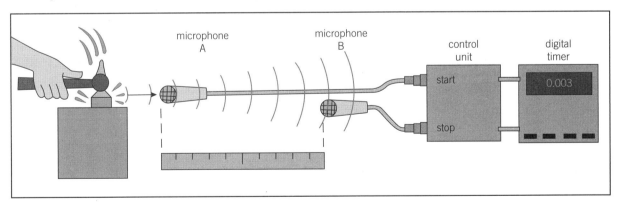

When the hammer hits the metal plate, microphone A receives the sound wave almost instantaneously. The sound travels a distance of 1.0 m (measured with a ruler) before it reaches microphone B. The digital timer starts when the sound wave reaches microphone A and stops when the sound wave reaches microphone B. The experiment is repeated several times and an average time found. The speed can then be found:

$$\text{Speed} = \frac{\text{distance}}{\text{average time}}$$
$$= \frac{1.0}{0.003}$$
$$= 330 \text{ m/s}$$

Echoes

Sound waves obey the same laws of reflection as light waves. Sound waves reflect from any surface. A reflected sound wave is called an **echo**.

Sonar is a method of measuring distances with sound. If a sound wave is sent from a boat and reflects off the sea-bed, it will be received a short time later. The depth of the sea-bed can then be calculated. Ultrasound images are built up in the same way.

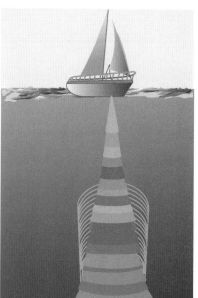

For example:

time between sending the sound wave and receiving the echo = 0.20 s
speed of sound in water = 1500 m/s
distance to the sea-bed $= \frac{0.20 \times 1500}{2} = 150$ m
(÷2 since 0.20 s is the time to travel to the sea-bed **and back**)

Examination style questions

1. The diagram illustrates a sound wave travelling through the air.

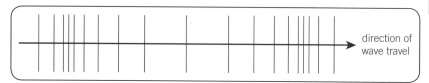

direction of wave travel

    a. On a copy of the diagram, mark clearly the direction in which the air particles are moving.
    b. Use the diagram to measure the wavelength of the sound wave in centimetres.
    c. The pitch of the sound wave is raised.

       State how the sound wave pattern would differ from that shown in the diagram above.

Cambridge IGCSE Physics 0625 Paper 22 Q6a June 2010

2. A square wooden block is made to rotate 2000 times per minute. A springy metal strip presses against the block, as shown in the diagram. A person nearby observes what is happening.

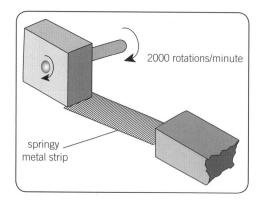

2000 rotations/minute

springy metal strip

    a. Calculate how many times per second the block rotates.
    b. Calculate the frequency of the sound caused by this arrangement.
    c. State whether or not this sound could be heard by the person nearby, and give a reason for your answer.

Cambridge IGCSE Physics 0625 Paper 2 Q8 June 2007

3. A man is using an axe to chop down a tree, as shown below.

a. A short time after the axe hits the tree, the man hears a clear echo.
    He estimates that the echo is heard 3 seconds after the axe hits the tree.
    i)   Suggest what type of obstacle might have caused such a clear echo.
    ii) The speed of sound in air is 320 m/s.
         Calculate the distance of the obstacle from the tree.

b. A branch from the tree falls into some shallow water in a pond nearby. The branch
    sets up a wave. The wave moves to the left a distance of 3.0 m before hitting the
    side of a moored boat and reflecting back again.

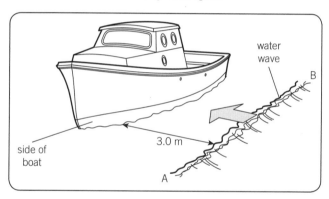

The wave takes 5.0 s to travel from AB to the boat and back to AB.
Calculate the speed of the water-wave.

Cambridge IGCSE Physics 0625 Paper 2 Q7 November 2006

Summary questions on Unit 3

1. Copy and fill in the blanks

There are two type of wave: _____ and _____. In a _____ wave the oscillations are perpendicular to the direction of travel of the _____. In a _____ wave the _____ are _____ to the direction of travel. The _____ of a wave is the distance between adjacent points on the wave that are at the same stage in their oscillation. The _____ _____ of a wave is the time for one complete wave. The frequency is the number of _____ passing a point each _____. Velocity (or speed), frequency and wavelength are related by the equation _____. When waves meet a barrier, they bounce off. This is called _____. When waves speed up or slow down the change direction. This is called _____. When waves pass through a gap in a barrier, they spread out. This is called _____.

2. Mark amplitude and wavelength on a copy of the diagram below.

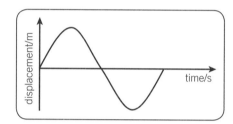

3. Copy and complete the following diagrams to show the reflected and refracted rays.

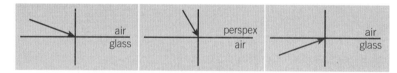

4. In each of the following cases, draw a diagram to show the incident, refracted and reflected rays and the normal to the boundary. Calculate the angle of **refraction** and state the angle of **reflection** in each case. The refractive index for air to glass is 1.5. The refractive index for glass to air is 0.67.
   a. Light incident at an angle of 56° to the normal in air on a glass surface.
   b. Light incident in glass on an air/glass boundary at an angle of 35° to the normal.
   c. Light incident in glass at an angle of 60° to the normal at an air/glass boundary.

   What is different about part c?

5. Match the terms on the right with the correct definition from the left hand column.

Ratio of speed of light in air to speed in material	Diffraction
Wave spreading out as it passes through a gap	Refractive index
Wave changing direction as its speed changes	Dispersion
Light bouncing off a smooth surface	Reflection
White light splitting into a spectrum as it is deviated by a prism	Refraction

6. On graph paper, draw the following ray diagrams to scale:

a. An object of height 2 cm at a distance of 5 cm from a convex lens of focal length 3 cm.

b. An object of height 3 cm at a distance of 5.5 cm from a convex lens of focal length 2.5 cm.

c. An object of height 2 cm at a distance of 2 cm from a convex lens of focal length 3 cm.

In each case, state a possible use for the arrangement.

7. a. What is the wavelength in air of electromagnetic radiation of frequency 1.0×10^{16} Hz? Which region of the spectrum does this belong to?

b. What is the frequency of the electromagnetic radiation which has a wavelength of 0.1 m in air? Which region of the spectrum does this belong to?

c. What is the wavelength in air of electromagnetic radiation of frequency 1.0×10^{20} Hz? Which region of the spectrum does this belong to?

8. Crossword

Across:

2 Region of the electromagnetic spectrum that has the highest energy (5)

4 The colour in the visible spectrum that is deviated the most by a prism (6)

6 Region between visible and X-rays in the electromagnetic spectrum (5-6)

8 Perpendicular to a boundary between two media (6)

12 The distance between the mid position of a particle on a wave and its maximum displacement (9)

13 Angle of incidence when the angle of refraction is 90° (8, 5)

17 Region of the electromagnetic spectrum with longer wavelengths than visible light (5-3)

18 This type of wave cannot travel in a vacuum (5)

19 Region of the electromagnetic spectrum with the longest wavelength (5)

20 20 to 20 000 is the _____ range that the human ear can hear. (9)

21 Speed = frequency x _____ (10)

Down:

1 Water waves do this when they pass from deep to shallow water (7)

3 Electromagnetic radiation used to cook food quickly (9)

5 Distance between the centre of a lens and its principal focus (5,6)

7 An image that cannot be focused on a screen (7)

9 Carries information by total internal reflection of light (7, 5)

10 The unit of frequency (5)

11 Region of a sound wave where the pressure is lowest (11)

14 Region of a sound wave where the pressure is highest (11)

15 Sound waves do this when they hit a hard smooth surface (7)

16 Lens that is thicker in the middle than the edges (6)

9. Explain what is meant by "dispersion".

10. Draw a diagram to show what happens to water waves, produced by a straight vibrating rod, when they are incident on a gap in a barrier.

11. A sound wave travels at 330 m/s in air with a wavelength of 0.1 m. What is its frequency? The wave enters water and the wavelength changes to 0.45 m. What is the speed of the sound wave in water?

12. A sound wave travels through a 2 m long metal rod in 0.42 ms. What is the speed of the wave?

Examination style questions on Unit 3

1. a. A ray of light passes through one surface of a glass prism at right angles to the surface, as shown in Figure 1.

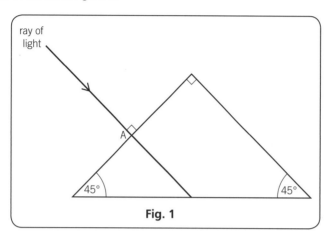

Fig. 1

   i) State why the ray is not deviated as it passes through the surface into the glass at A.

   ii) On a copy of the diagram, use a ruler to help you draw the rest of the path of the ray, until it has emerged again into the air.

   b. Figure 2 shows a periscope that uses two plane mirrors.

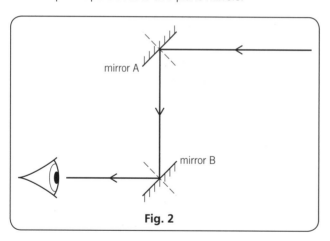

Fig. 2

   i) On a copy of the diagram, clearly mark the angle of incidence *i* and the angle of reflection *r* at mirror A.

   ii) State the equation linking *i* and *r*.

   iii) Use a ruler to redraw the periscope, but using prisms like that in Figure 1 instead of mirrors at A and B.

Cambridge IGCSE Physics 0625 Paper 22 Q7 June 2010

2. The speed of sound in air is 332 m/s. A man stands 249 m from a large flat wall, as shown below, and claps his hands once.

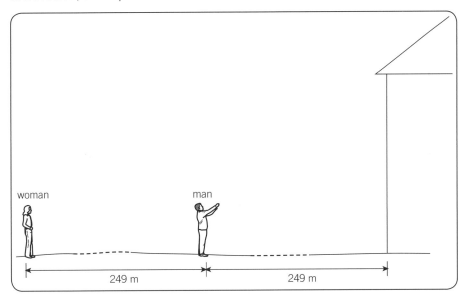

    a.   Calculate the interval between the time when the man claps his hands and the time when he hears the echo from the wall.

    b.   A woman is standing 249 m further away from the wall than the man. She hears the clap twice, once directly and once after reflection from the wall.
How long after the man claps does she hear these two sounds? Pick **two**.

      0.75 s   1.50 s   2.25 s   3.00 s

Cambridge IGCSE Physics 0625 Paper 2 Q9 June 2005

3. Figures 1 and 2 show wavefronts of light approaching a plane mirror and a rectangular glass block, respectively.

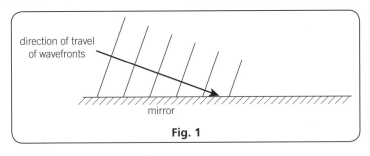

Fig. 1

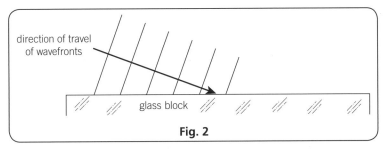

Fig. 2

    a.   On copies of the diagrams draw wavefronts to show what happens after the waves strike the surface.

    b.   In Figure 2, the waves approaching the block have a speed of 3.0×10^8 m/s and an angle of incidence of 70°. The refractive index of the glass of the block is 1.5.

      i)   Calculate the speed of light waves in the block.

      ii)   Calculate the angle of refraction in the block.

Cambridge IGCSE Physics 0625 Paper 3 Q7 June 2008

Extended

4. a. The diagram below shows two rays of light from a point O on an object. These rays are incident on a plane mirror.

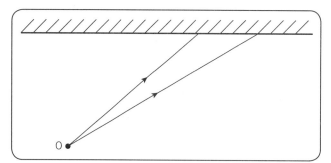

  i) Copy the diagram and continue the paths of the two rays after they reach the mirror. Hence locate the image of the object.O. Label the image I.
  ii) Describe the nature of the image I.

Cambridge IGCSE Physics 0625 Paper 3 Q7 November 2005

5. Two students are asked to determine the speed of sound in air on the school playing fields.
  a. List the apparatus they need.
  b. List the readings that the students need to take.
  c. State how the speed of sound is calculated from the readings.
  d. State one precaution that could be taken to improve the accuracy of the value obtained.
  e. The table gives some speeds.

speed/m/s	speed of sound in air	speed of sound in water
10		
100		
1000		
10 000		

  Copy the table and place a tick to show the speed which is closest to
  i) the speed of sound in air,
  ii) the speed of sound in water.

Cambridge IGCSE Physics 0625 Paper 3 Q7 June 2007

Extended

6. The diagram to the right shows white light incident at P on a glass prism. Only the refracted red ray PQ is shown in the prism.
  a. Copy the diagram, draw rays to complete the path of the red ray and the whole path of the violet ray up to the point where they hit the screen. Label the violet ray.
  b. The angle of incidence of the white light is increased to 40°. The refractive index of the glass for the red light is 1.52. Calculate the angle of refraction at P for the red light.
  c. State the approximate speed of
  i) the white light incident at P,
  ii) the red light after it leaves the prism at Q.

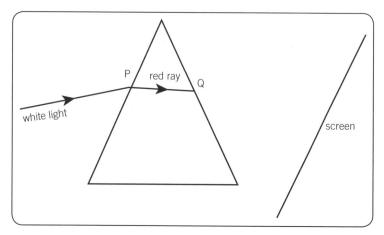

Cambridge IGCSE Physics 0625 Paper 3 Q6 June 2006

4 Electricity and magnetism

4.1 Magnetism

A north magnetic pole is actually a north seeking pole, i.e. if it is free to rotate, it will point towards the Earth's north pole.

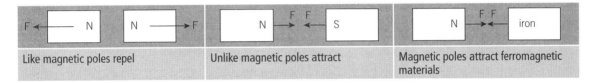

Like magnetic poles repel	Unlike magnetic poles attract	Magnetic poles attract ferromagnetic materials

Ferromagnetism

Iron is a ferromagnetic material. It contains what we can think of as tiny magnets called **domains**. In an unmagnetised piece of iron, the domains are arranged randomly, but when a piece of iron is placed near a magnet, the domains line up because they are all attracted by the magnet.

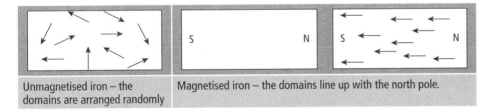

Unmagnetised iron – the domains are arranged randomly	Magnetised iron – the domains line up with the north pole.

This table shows how a magnet can pick up an unmagnetised piece of iron. When the domains line up, a south pole is formed opposite the north pole of the magnet. The poles attract. This is **induced** magnetism. When the magnet is removed, the piece of iron will quickly lose its magnetism. The domains become randomly arranged again and the **temporary magnetism** is lost.

Steel is made from iron and carbon and so it is also **ferromagnetic**. If steel is used instead of iron, the steel will keep become less strongly magnetised, but it will **retain** some of its induced magnetism and become a **permanent magnet**. It will remain magnetised until it is banged on the table, hammered, heated or placed in a coil with an alternating current flowing in it.

Field around a bar magnet

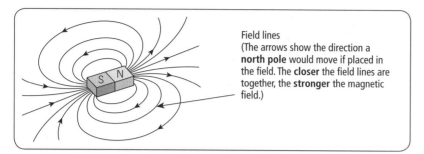

Field lines
(The arrows show the direction a **north pole** would move if placed in the field. The **closer** the field lines are together, the **stronger** the magnetic field.)

If small plotting compasses are placed around the bar magnet, the compasses show the direction of the magnetic field. If small pieces of iron (iron filings) are sprinkled around the magnet they too will line up with the magnetic field and show the field lines.

Electromagnetism

By convention, current is always shown as flowing from the positive side of a cell to the negative side in a complete circuit. **Conventional current is the flow of positive charge**.

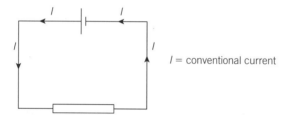

I = conventional current

Electrons are **negatively** charged, which means that they flow in the **opposite direction** to conventional current.

Current is measured using an **ammeter**, which must be placed in **series** in a circuit.

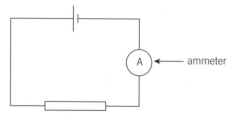

ammeter

When a current flows through a wire, a magnetic field is produced around the wire. The direction of the magnetic field depends on the direction of the current (from + to – around the circuit) and is given by the right hand grip rule as shown below.

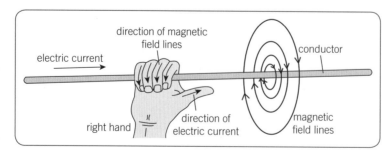

The **field lines** point in the direction of the **fingers** of your right hand when your **thumb** points in the direction of the **current**.

If the current flows in the opposite direction, then you must turn your hand round and see that the field lines now go round the wire in the opposite direction.

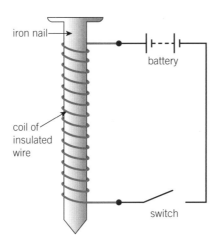

When a current flows through a coil of wire, a magnetic field is created around and inside the coil. The pattern of the **field lines** outside the coil is identical to a **bar magnet**. Objects made of ferromagnetic materials can be magnetised by placing them inside the coil.

An electromagnet can be switched off by opening the switch and cutting off the current from the battery. Its strength can be increased by increasing the voltage of the supply.

Examination style questions

1. a. Four rods are shown in Figure 1.

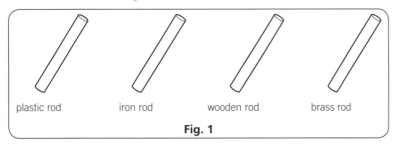

Fig. 1

State which of these could be held in the hand at one end and be

   i)   magnetised by stroking it with a magnet,
   ii) charged by stroking it with a dry cloth.

   b. Magnets A and B in Figure 2 are repelling each other.

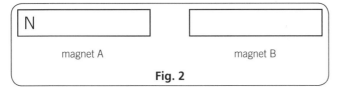

Fig. 2

The north pole has been labelled on magnet A.

On a copy of Figure 2, label the other three poles.

   c. Figure 3 shows a plotting compass with its needle pointing north.

Fig. 3

A brass rod is positioned in an east-west direction. A plotting compass is put at each end of the brass rod, as shown in Figure 4.

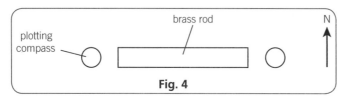

Fig. 4

On a copy of Figure 4, mark the position of the pointer on each of the two plotting compasses.

Cambridge IGCSE Physics 0625 Paper 2 Q8a, b&d June 2009

2. a. Two magnets are laid on a bench. End A of an unidentified rod is held in turn above one end of each magnet, with the results shown below.

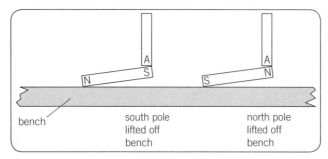

i) Suggest what the unidentified rod is made from.
ii) State what, if anything, happens when the end A is held over one end of
   1. an unmagnetised iron bar,
   2. an uncharged plastic rod.

b. The diagram below shows four identical plotting compasses placed around a bar magnet where the magnetic field of the surroundings can be ignored. The pointer has only been drawn on one plotting compass.

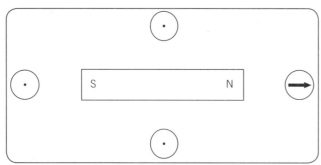

Copy the diagram, draw the pointers on the other three plotting compasses to indicate the directions of the magnetic field of the bar magnet in those three places.

Cambridge IGCSE Physics 0625 Paper 2 Q8 November 2005

3. The diagram below shows a bar magnet on a board in a region where the magnetic field of the surroundings is so weak it can be ignored. The letters N and S show the positions of the north and south poles of the magnet. Also on the diagram are marked four dots.

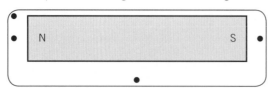

a. Copy the diagram and carefully draw four magnetic field lines, one passing through each of the four dots. The lines you draw should begin and end either on the magnet or at the edge of the board.
b. On one of your lines, put an arrow to show the direction of the magnetic field.

Cambridge IGCSE Physics 0625 Paper 2 Q11 June 2007

4.2 Electrical quantities

Electric charge

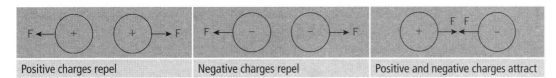

| Positive charges repel | Negative charges repel | Positive and negative charges attract |

The region around an electric charge where another charge experiences a force is called an **electric field**. The field lines show the direction a positive charge would move if placed in the field.

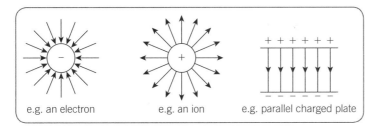

▲ Electric field lines for three different examples

An object will become charged negative if it gains electrons and positive if it loses electrons.

Charging by induction

1.

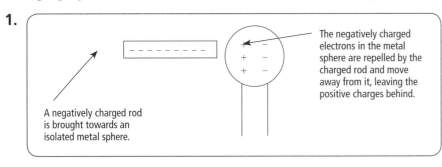

A negatively charged rod is brought towards an isolated metal sphere.

The negatively charged electrons in the metal sphere are repelled by the charged rod and move away from it, leaving the positive charges behind.

2.

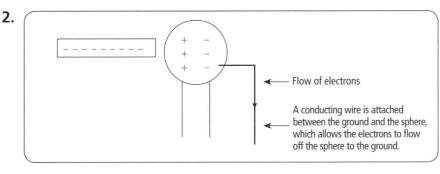

Flow of electrons

A conducting wire is attached between the ground and the sphere, which allows the electrons to flow off the sphere to the ground.

Extended

Extended

3.

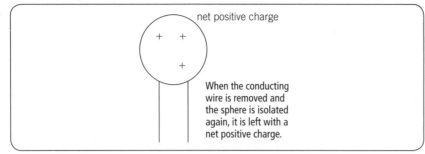

net positive charge

When the conducting wire is removed and the sphere is isolated again, it is left with a net positive charge.

Electric current

Charge is measured in **coulombs (C)**. Good **conductors**, such as metals, have **electrons** that are free to move through their structure. Poor conductors (**insulators**) do not have freely moving charged particles within their structure. Examples of insulators are rubber, plastic and paper.

The flow of charge through a conductor is called the **current**. Current will only flow through a conductor if there is a **potential difference** (PD) between the ends of the conductor. Conventional current flow is from the positive side of the power supply to the negative. In metals, electrons flow in the opposite direction.

The current in amperes is equal to the charge in coulombs passing a point every second.

Charge (C) = current (A) × time (s)

$$Q = I \times t$$

Worked examples

1. A charge of 0.01 C passes a point in a circuit every 0.2 s. What is the current flowing?

2. How long does it take for a charge of 30 C to pass a point in a circuit when a current of 0.8 A flows?

3. How much charge will pass a point in a circuit when a current of 0.5 A flows for 1 minute?

Answers

1. $I = \dfrac{Q}{t}$

$= \dfrac{0.01}{0.2}$

$= 0.05$ A

2. $t = \dfrac{Q}{I}$

$= \dfrac{30}{0.8}$

$= 37.5$ s

3. $Q = I \times t$

$= 0.5 \times 60$

$= 30$ C

Note: In question 3, 1 minute must be changed into 60 seconds.

Electromotive force and potential difference

An electrical supply (a power pack, cell or battery) provides electrical energy, which is carried round a circuit by the current. The electromotive force or **e.m.f.** of a supply is the energy per coulomb of charge, it is measured in **volts (V)**.

Potential difference or **voltage** across a component in a circuit is the energy required per coulomb of charge to drive the current through that component. It is measured in **volts (V)**.

The potential difference (**PD**) across a component is measured using a **voltmeter** in parallel with the component.

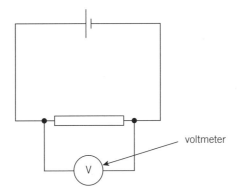

voltmeter

Resistance

Metals at a **constant temperature** have a constant resistance, measured in **ohms** (Ω).

$$\text{Resistance } (\Omega) = \frac{\text{potential difference (V)}}{\text{current (A)}}$$

$$R = \frac{V}{I}$$

For a constant potential difference, **increasing the resistance decreases the current**.

For a constant resistance, **increasing the potential difference increases the current**.

Worked examples

1. A potential difference of 20 V is required for a current of 0.5 A to flow through a resistor. What is its resistance?

2. A current of 0.01 A flows through a resister of 1 kΩ. What is the potential difference across the resistor?

3. How much current flows when a potential difference of 5 V is applied to a resistor of 10 Ω?

Answers

1. $R = \dfrac{V}{I}$

$\quad = \dfrac{20}{0.5}$

$\quad = 40 \ \Omega$

3. $I = \dfrac{V}{R}$

$\quad = \dfrac{5}{10}$

$\quad = 0.5 \ A$

2. $V = I \times R$

$\quad = 0.01 \times 1000$

$\quad = 10\,V$

Note: In question 2, kΩ must be changed into Ω. 1 kΩ=1000 Ω.

An experiment to find the resistance of an unknown resistor

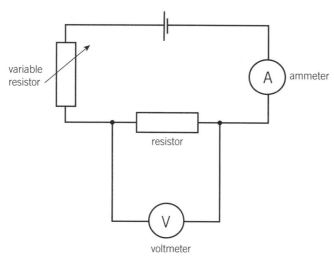

▲ Test circuit

Vary the potential difference across the unknown resistor by changing the resistance of the variable resistor. Measure the PD across the unknown resistor each time you change the resistance of the variable resistor, using the voltmeter and the corresponding values of current using the ammeter.

Results

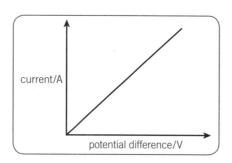

Analysis

$$\text{Resistance} = \frac{V}{I} = \frac{1}{\text{gradient of graph}}$$

Alternatively, resistance can be calculated by working out V/I for several values of V and I and an average resistance found.

The gradient of the graph is constant because the resistance is constant provided that the temperature of the resistor remains constant. Should the resistor become hot (as in the filament of a lamp), the resistance increases and the gradient of the current-voltage graph decreases.

The resistance of a wire

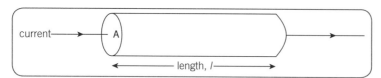

A metal wire has a length l and a cross-sectional area A. When the length of the wire is increased, the current has to travel further through the wire and so the resistance increases. When the cross-sectional area is increased by increasing the diameter of the wire, the current has a greater area to travel through and so the resistance decreases.

The resistance of the wire is directly proportional to the length of the wire and inversely proportional to the cross-sectional area.

Extended

Current – voltage characteristic of a filament lamp

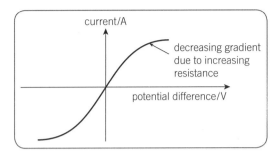

As potential difference increases, the current increases, which increases the temperature of the filament. The metal ions gain more energy and vibrate more, which increases the frequency of collisions between the moving electrons and the metal ions. Hence the resistance increases.

Electrical power

Electrical energy is transferred from the battery or power supply in a circuit to the components in the circuit by the electrons. The component transforms the electrical energy into other forms (for example a bulb converts electrical energy into heat and light). The rate at which the energy is transformed is the **power**. Power can be calculated from the formula:

Power (W) = potential difference (V) × current (A)

$$P = VI$$

Energy = power × time

so $E = VIt$

Worked examples

1. What is the power of a bulb that allows a current of 1.5 A to flow when there is a potential difference of 8 V across it?

2. What is the current flowing through a bulb of power 40 W when there is a potential difference of 200 V across it?

3. What potential difference is required to produce a current of 0.3 A in a bulb of power 60 W?

Answers

1. $P = V \times I$

   $= 8 \times 1.5$

   $= 12$ W

2. $I = \dfrac{P}{V}$

   $= \dfrac{40}{200}$

   $= 0.2$ A

3. $V = \dfrac{P}{I}$

   $= \dfrac{60}{0.3}$

   $= 200$ V

Examination style questions

1. The table below shows the potential difference (PD) needed at different times during a day to cause a current of 0.03 A in a particular thermistor.
   a. Calculate the two values missing from the table.

time of day (24-hour clock)	0900	1200	1500	1800
PD/V	15.0	9.9		7.5
resistance/Ω	500		210	250

b. Copy the axes below, plot the four resistance values given in the table.

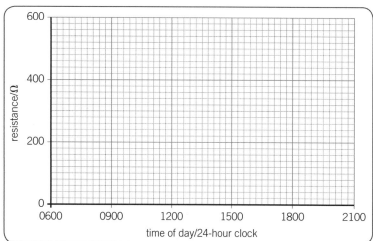

c. i) Draw a smooth curve through your points.
    ii) Why do we draw a smooth curve rather than a series of straight lines joining the points?
d. The thermistor is a circuit component with a resistance that decreases as the temperature increases.
    i) From your graph, estimate the time of day when the temperature was greatest.
    ii) State the reason for your answer to **d(i)**.

Cambridge IGCSE Physics 0625 Paper 2 Q6 June 2005

2. A length of bare uniform resistance wire is included in the circuit shown in the figure below. Contact C can be moved to any position along the resistance wire.

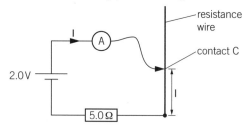

a. On the axes provided, sketch the graph that relates the current I in the circuit to the length l of the resistance wire.

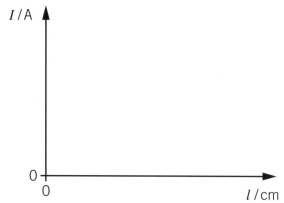

b. Calculate the reading on the ammeter when the length l is zero.
c. Contact C is moved so that the resistance of the length l of the resistance wire is 15.0 Ω.

   Calculate

   i) the total resistance of the circuit,
   ii) the new ammeter reading.
d. When l = 25 cm, the reading on the ammeter is half that found in **(b)**.

   Calculate the resistance of 25 cm of the resistance wire.

e. Which of the following effects is caused by the current in the resistance wire?
    Tick the boxes alongside **two** correct effects.

    heating   ☐

    light   ☐

    sound   ☐

    magnetism   ☐

Cambridge IGCSE Physics 0625 Paper 22 Q10 June 2012

3. A student has a power supply, a resistor, a voltmeter, an ammeter and a variable resistor.

a. The student obtains five sets of readings from which he determines an average
    value for the resistance of the resistor.

    Draw a labelled diagram of a circuit that he could use.

b. Describe how the circuit should be used to obtain the five sets of readings.

c. The following circuit is set up and the reading on the ammeter is 0.5 A.

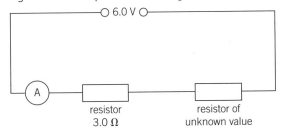

i)   Calculate the value of the unknown resistor.
ii) Calculate the charge passing through the 3.0 Ω resistor in 120 s.
iii) Calculate the power dissipated in the 3.0 Ω resistor.

Cambridge IGCSE Physics 0625 Paper 3 Q8 June 2005

4. a. In the figure below, S is a metal sphere standing on an insulating base. R is a
        negatively charged rod placed close to S.

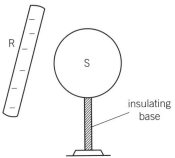

i)   Name the particles in S that move when R is brought close to S.
ii) Copy the figure and add + signs and – signs to suggest the result of this
     movement.
iii) Describe the actions which now need to take place so that S becomes positively
     charged with the charge distributed evenly over its surface. A positively charged
     object is **not** available.

b. During a thunderstorm, the potential difference between thunderclouds and the
    ground builds up to 1.5×10^6 V. In each stroke of lightning, 30 C of charge passes
    between the thunderclouds and the ground. Lightning strokes to the ground occur,
    on average, at 2 minute intervals.

    Calculate
    i)   the average current between the thunderclouds and the ground,
    ii) the energy transferred in each stroke of lightning.

Cambridge IGCSE Physics 0625 Paper 31 Q8 June 2012

Extended

Practical question

A student uses the circuit shown below to investigate the resistance of a piece of wire.

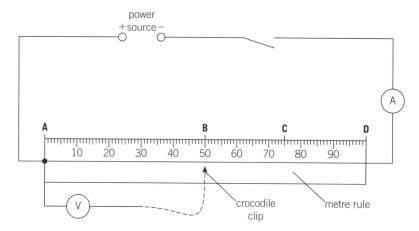

The student measures the current *I* in the wire. She then measures the p.d. *V* across **AB**, **AC** and **AD**.

The student's readings are shown in the table below.

section of wire	*l*/cm	*I*/A	*V*/V	*R* /
AB		0.375	0.95	
AC		0.375	1.50	
AD		0.375	1.95	

1. Using the diagram, record in the table the length *l* of each section of wire.

2. Show the positions of the pointers of the ammeter reading 0.375 A, and the voltmeter reading 1.50 V on copies of the blank ammeter and voltmeter below.

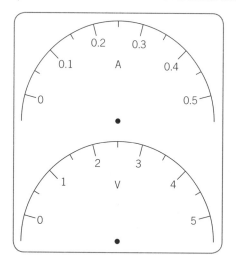

3. Calculate the resistance *R* of the sections of wire **AB**, **AC** and **AD** using the equation $R = \dfrac{V}{I}$.

4. Record these values of *R*, to a suitable number of significant figures, in the table.

5. Complete the column **heading** for the *R* column of the table.

6. Use your results to predict the resistance of a 1.50 m length of the same wire. Show your working.

Cambridge IGCSE Physics 0625 Paper 6 Q3 June 2005

4.3 Electric circuits

KEY IDEAS

✓ Current is the same in each part of a series circuit, but is shared in a parallel circuit
✓ PD is the same across each part of a parallel circuit, but is shared across a series circuit
✓ Thermistors and LDRs can be used as sensors
✓ Digital signals have only two states: high and low
✓ Logic gates can be used to process signals to produce a desired output

Symbols for electrical components

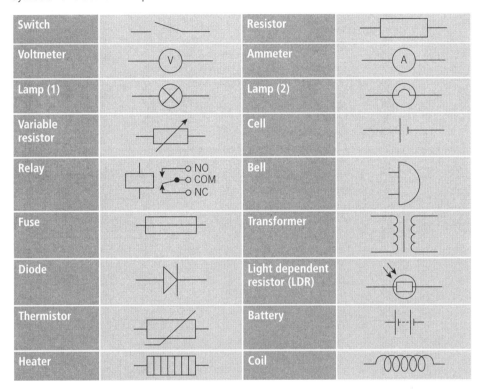

Drawing circuit diagrams

Use a pencil and a ruler to draw the connecting wires.
Do not draw the connecting wires through the components i.e.
Always use the correct symbols for the components shown in the table above.

The e.m.f of a cell

E.m.f stands for electromotive force. The e.m.f of a cell is the maximum PD it can provide.

Adding cells in series

When cells are arranged in series the total e.m.f is equal to the sum of the electromotive forces of the cells.

Series and parallel circuits

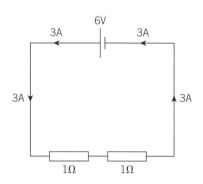

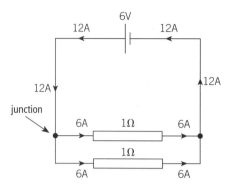

▲ Series circuit   ▲ Parallel circuit

Series circuit	Parallel circuit
In a series circuit the current is the **same** all the way round the circuit.	In a parallel circuit the current splits at the junction and is **shared** between the resistors ie 12A = 6A + 6A
Adding resistors in series **increases** the total resistance in the circuit.	Adding resistors in parallel **decreases** the total resistance of the circuit.
In a series circuit, the total resistance is simply the sum of the individual resistances. In the example above, **total resistance = 1 Ω + 1 Ω = 2 Ω**	In a parallel circuit, the total resistance is given by the formula $$\frac{1}{\text{total resistance}} = \frac{1}{R_1} + \frac{1}{R_2}$$ In the example above, $$\frac{1}{\text{total resistance}} = \frac{1}{1} + \frac{1}{1}$$ $$= 2$$ **total resistance = $\frac{1}{2}$ = 0.5 Ω** Note that the total resistance is less than each of the resistors.
The potential difference across each resistor can be calculated using $V = IR$. In the example above, $V = 3 \times 1 = 3$ V, so each resistor has a PD of **3 V** across it.	The potential difference across each resistor can be calculated using $V = IR$. In the example above $V = 6 \times 1 = 6$ V, so each resistor has a PD of **6 V** across it.
In a series circuit, the potential difference is **shared** between the resistors. This is because the energy from the cell is shared between the resistors.	In a parallel circuit, the potential difference across each resistor is the **same** as the potential difference across the cell. This is because after picking up energy from the cell a charge only passes through one of the resistors (not both).
If one of the resistors broke, the circuit would be broken and **no current** would flow.	If one of the resistors broke, the **current could still flow** through the second resistor, although the current would be smaller because there would now be a greater total resistance in the circuit.

Strings of party lights used to be wired in series, which meant if one of the bulbs broke, all of the bulbs would go out. Now, they are wired in parallel so that the rest of the bulbs will still work if one breaks.

Adding bulbs in parallel decreases the total resistance of the circuit.

Action and use of circuit components

Relay

A relay is an electromagnetic switch that often uses a small current to switch on a much larger current, which may be dangerous if done directly.

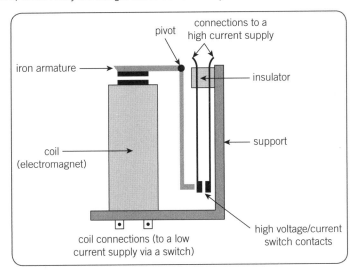

When the switch is closed, a small current flows through the coil, creating a magnetic field. The iron core of the coil becomes magnetised and attracts the iron armature, which pivots and pushes the contacts together. This closes the high current circuit and switches on heavy machinery, for example. The advantage of this arrangement is that the operator cannot come in contact with the high current supply.

Potentiometer

A potentiometer can be made from a **variable resistor**. Sliding the moving contact along the length of the resistor moves the contact represented by the arrow in the diagram. This changes the voltage on the voltmeter. Increasing the resistance in parallel with the voltmeter, increases the share of the potential difference from the cell, and the reading on the voltmeter increases.

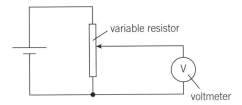

If the voltmeter is replaced with another component, the potential difference across the component can be varied with the potentiometer between 0 V and the maximum PD from the cell.

Thermistor

The thermistor is a component whose **resistance decreases as the temperature increases**. This means that it can be used as a **temperature sensor**.

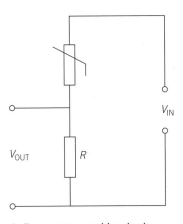

▲ Temperature sensitive circuit

When the thermistor is warmed, its resistance decreases and it takes a smaller share of the potential difference from the power supply. *R* takes a larger share of the supply PD.

This circuit could be used to turn on a warning light to show that the hob of an electric cooker is hot, or as an indicator light to show that hair straighteners have reached their operating temperature.

Light dependent resistor (LDR)

The light dependent resistor (LDR) is a component whose **resistance decreases as the light intensity increases**. This means that it can be used as a light sensor.

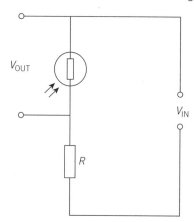

▲ Light sensitive circuit

When the light intensity on the LDR decreases, its resistance increases and it takes a larger share of the potential difference from the power supply. Resistor *R* takes a smaller share of the potential difference from the power supply.

This circuit could be used to switch on a light at night.

Diode

A diode only allows current to flow **one way** through it, the direction in which the **arrow** is pointing.

This property of the diode is used in the conversion of a.c. current to d.c. Alternating current (a.c.) repeatedly changes the direction in which it flows around the circuit. Since the diode only allows current to flow in one direction, it changes the alternating current to direct current. This is called **rectification**.

Extended

4.4 Digital electronics

A digital system consists of an input sensor (such as a simple push switch) and a processor circuit, which controls the voltage to an output device (such as an electric bell). The processor circuit consists of a series of logic gates. Logic gates respond to small voltages, which are either on or off (digital signals). They do not respond to analogue signals.

An analogue signal (V) varies continuously in **amplitude**.	
A digital signal (V) has only two states: high and low (or on and off, or 1 and 0)	

Logic gates

Logic gates are circuits containing transistors and other components. They transform a digital input voltage into an output, which depends on the type of logic gate. The input and output voltages are given as 1 or 0 (on or off) and can be represented in a truth table.

Logic gate	Symbol	Truth table
NOT	A —▷o— Y	A | Y 1 | 0 0 | 1
AND	A, B —D— Y	A | B | Y 0 | 0 | 0 1 | 0 | 0 0 | 1 | 0 1 | 1 | 1
OR	A, B —D— Y	A | B | Y 0 | 0 | 0 1 | 0 | 1 0 | 1 | 1 1 | 1 | 1
NAND	A, B —Do— Y	A | B | Y 0 | 0 | 1 1 | 0 | 1 0 | 1 | 1 1 | 1 | 0
NOR	A, B —Do— Y	A | B | Y 0 | 0 | 1 1 | 0 | 0 0 | 1 | 0 1 | 1 | 0

A NOT gate gives an output that is the opposite of the input.
An AND gate only gives an output if the input A *and* input B are both 1.
An OR gate gives an output if input A *or* input B is 1.

Logic gates can be combined to perform different functions in electrical circuits.

The truth table for the arrangement on the left is as follows:

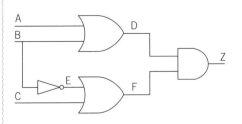

A	B	C	D	E	F	Z
0	0	0	0	1	1	0
1	0	0	1	1	1	1
0	1	0	1	0	0	0
0	0	1	0	1	1	0
1	1	0	1	0	0	0
1	0	1	1	1	1	1
0	1	1	1	0	1	1
1	1	1	1	0	1	1

Extended

This is a very complicated logic gate circuit! It has three inputs A, B and C and a final output Z. Start by reading columns A + B and looking at the output from the OR gate D. If either A or B or both are high (1), then D is high (1). Now check the input B and the output E from the NOT gate. The output from E is the opposite to the input at B. E + C are the inputs to an OR gate whose output is F. Finally, look at the inputs D + F and the output Z to the AND gate. Check that both D and F must be high (1) for Z to be high (1).

The following block diagram shows a circuit that will switch on a warning light at night when the temperature is too low.

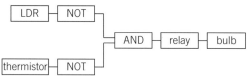

At night, the LDR output is 0, which is reversed by the NOT gate.
In the cold the thermistor output is 0, which is changed to 1 by the NOT gate.
This switches at the AND gate. The current is increased by the relay and the bulb lights.

Examination style questions

1. A student has two wires A and B. She connects each in turn between the terminals P and Q in the circuit as shown in the figure.

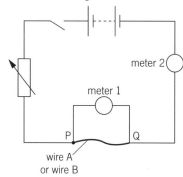

   For each wire, she measures the current in the wire when there are various potential differences across it.

   a. i)   Which of the two meters measures the current?
       ii) Name this type of meter.
   b. i)   Which of the two meters measures the potential difference (p.d.)?
       ii) Name this type of meter.
   c. When the student draws the graphs of p.d. against current for the two wires, she gets the lines shown in the graph.

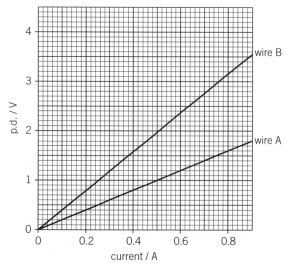

i)   From the graph shown, find the p.d. across wire A when there is a current of 0.8 A in it.

ii) Calculate the resistance of wire A when the current in it is 0.8 A.

iii) From the graph, how can you tell that the resistance of wire A remains constant during the experiment?

iv) How can you tell that the resistance of wire B is greater than the resistance of wire A?

v)   Wires A and B are made of the same material and have the same thickness.

State, giving your reasons, which of the wires is the longer wire.

Cambridge IGCSE Physics 0625 Paper 22 Q8 November 2012

2. a. State the electrical quantity that has the same value for each of two resistors connected to a battery

i)   when they are in series,

ii) when they are in parallel.

b. The figure shows a circuit with a 1.2 kΩ resistor and a thermistor in series. There is no current in the voltmeter.

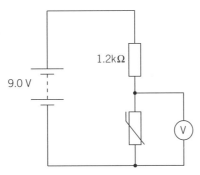

Calculate the voltmeter reading when the resistance of the thermistor is 3.6 kΩ.

c. Figure 2 shows a fire-alarm circuit. The circuit is designed to close switch S and ring bell B if there is a fire.

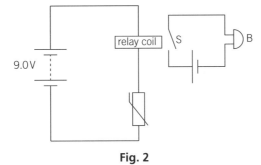

Fig. 2

Explain the operation of the circuit.

Cambridge IGCSE Physics 0625 Paper 31 Q10 November 2012

3. a. A coil of wire is connected into a circuit containing a variable resistor and a battery. The variable resistor is adjusted until the potential difference across the coil is 1.8 V. In this condition, the current in the circuit is 0.45 A.

Calculate

i)   the resistance of the coil,

ii) the thermal energy released from this coil in 9 minutes.

b. The coil in part **(a)** is replaced by one made of wire which has half the diameter of that in **(a)**.

When the potential difference across the coil is again adjusted to 1.8 V, the current is only 0.30 A.

Calculate how the length of wire in the second coil compares with the length of wire in the first coil.

Cambridge IGCSE Physics 0625 Paper 31 Q11 June 2010

4. The diagram below shows a low-voltage lighting circuit.

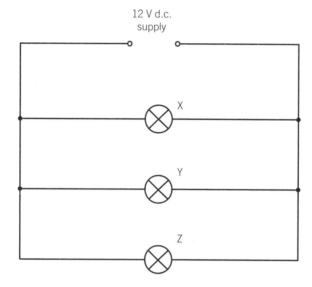

a. Copy the diagram, indicate with a dot and the letter S, a point in the circuit where a switch could be placed that would turn off lamps Y and Z at the same time but would leave lamp X still lit.
b. i) Draw the circuit symbol for a component that would vary the brightness of lamp X.
   ii) On your diagram, mark with a dot and the letter R where this component should be placed.
c. Calculate the current in lamp Y.
d. The current in lamp Z is 3.0 A. Calculate the resistance of this lamp.
e. The lamp Y is removed.
   i) Why do lamps X and Z still work normally?
   ii) The current in lamp X is 1.0 A. Calculate the current supplied by the battery with lamp Y removed.

Cambridge IGCSE Physics 0625 Paper 3 Q8 November 2006

5. The circuit shown in Figure 1 uses a 12 V battery.

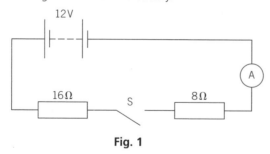

Fig. 1

a. Switch S is open, as shown.

State the value of
   i) the reading on the ammeter,
   ii) the potential difference (p.d.) across S.

b. Switch S is now closed.
   i) Calculate the current in the ammeter.
   ii) Calculate the p.d. across the 8 Ω resistor.
c. The two resistors are now connected in parallel.

Calculate the new reading on the ammeter when S is closed, stating clearly any equations that you use.

Cambridge IGCSE Physics 0625 Paper 31 Q10 June 2009

6. The diagram below shows a high-voltage supply connected across two metal plates.

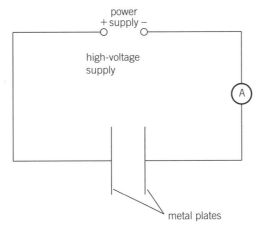

When the supply is switched on, an electric field is present between the plates.
a. Explain what is meant by an *electric field*.
b. Copy the diagram, draw the electric field lines between the plates and indicate their direction by arrows.
c. The metal plates are now joined by a high-resistance wire. A charge of 0.060 C passes along the wire in 30 s.
   Calculate the reading on the ammeter.
d. The potential difference of the supply is re-set to 1500 V and the ammeter reading changes to 0.0080 A. Calculate the energy supplied in 10 s. Show your working.

Cambridge IGCSE Physics 0625 Paper 3 Q8 November 2005

7. a. Draw the symbol for a NOR gate. Label the inputs and the output.
   b. State whether the output of a NOR gate will be high (ON) or low (OFF) when
      i) one input is high and one input is low,
      ii) both inputs are high.
   c. The diagram below shows a digital circuit made from three NOT gates and one NAND gate.

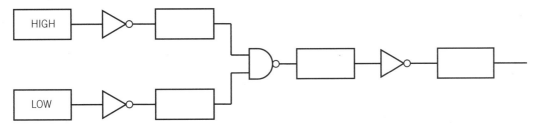

   i) Write HIGH or LOW in each of the boxes on the diagram
   ii) State the effect on the output of changing both of the inputs.

Cambridge IGCSE Physics 0625 Paper 3 Q9 November 2005

Practical question

A student is investigating the relationship between potential difference *V* across a resistor and the current *I* in it. The diagram below shows the apparatus that the student is using.

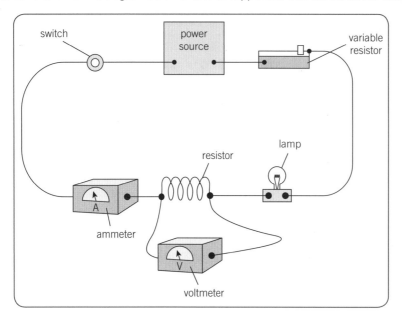

1. Draw the circuit diagram of the circuit shown above. Use standard circuit symbols.

2. The student is using a lamp to show when the current is switched on.
   Why is it unnecessary to use the lamp?

3. State which piece of apparatus in the circuit is used to control the size of the current.

4. The student removes the lamp from the circuit. She is told that the resistance of a conductor is constant if the temperature of the conductor is constant. She knows that the current in the resistor has a heating effect. Suggest two ways in which the student could minimise the heating effect of the current in the resistor.

5. The diagram below shows a variable resistor with the sliding contact in two different positions.

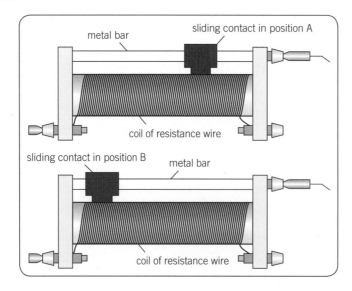

   State which position, A or B, shows the higher resistance setting. Explain your answer.

Adapted from Cambridge IGCSE Physics 0625 Paper 6 Q5 November 2006

4.5 Dangers of electricity

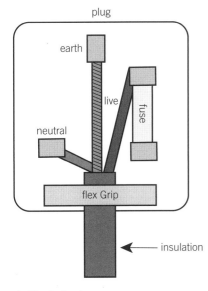

▲ The 3 pin plug

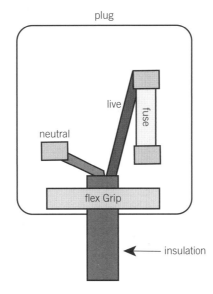

▲ The 2 pin plug

Safety features of the 3 pin plug and the 2 pin plug

- If the live wire comes loose and touches the metal casing of the appliance, the earth wire carries the current safely to ground and the fuse melts. This prevents electric shock.
- In the 2 pin plug the earth wire is carried in grooves in the plastic case.
- If the current in an appliance becomes too large, say because there is a short circuit, the thin wire inside the fuse melts and breaks the circuit.
- The plastic coating over the cable insulates the conducting wires.

Water conducts electric current and so you should never touch an electrical appliance with wet hands or operate electrical equipment in wet conditions.

In the home, circuit breakers protect us from electrical fires.

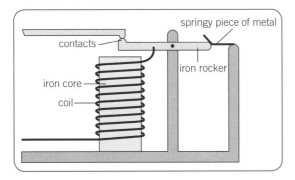

If the current flowing through the coil of wire becomes **too large**, the iron core becomes strongly magnetised and attracts the iron rocker with enough force to pull it down. This **opens the contacts** and **breaks the circuit** so that the current can no longer flow. The circuit breaker can be reset by flicking a switch, which pushes down the springy piece of metal and pulls the rocker back up to close the contacts.

A "blown" fuse has to be replaced, but a circuit breaker can easily be reset by lifting the iron rocker back into position.

Examination style question

1. For each hazard, draw a line to the appropriate protection.

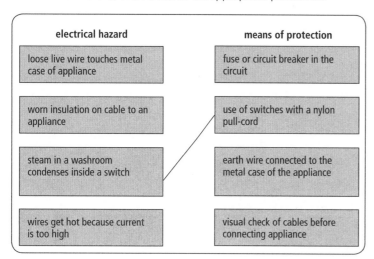

Cambridge IGCSE Physics 0625 Paper 2 Q12 June 2006

4.6 Electromagnetic effects

KEY IDEAS

✓ When a coil cuts magnetic field lines, an e.m.f. is induced across the coil; this principle is used in the a.c. generator

✓ A transformer can be used to step up and step down a.c. voltage

✓ The direction of the force on a current carrying conductor in a magnetic field is given by the left hand rule

✓ A coil of wire carrying a current in a magnetic field will turn due to the forces on it; this principle is used in the d.c. motor.

Electromagnetic induction

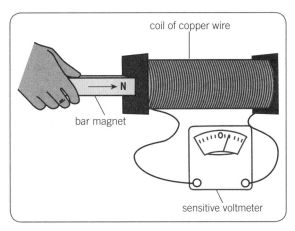

coil of copper wire

bar magnet

sensitive voltmeter

When the north pole of the bar magnet is moved into the coil, the needle on the sensitive voltmeter briefly moves to the right before returning to the centre.

The movement of the magnetic field lines due to the bar magnet **induces** an e.m.f. across the coil, which is then measured by the voltmeter. This is because as the magnet moves into the coil, the coil **cuts the magnetic field lines**. The magnetic field linking with the coil is changing.

When the north pole is moved out of the coil, the needle briefly moves to the left. This is because the coil is cutting the field lines in the opposite direction and so the **induced e.m.f.** is in the opposite direction.

The voltmeter can also be made to briefly move to the left if a south pole is moved into the coil.

Ways to increase the induced e.m.f.:

- Move the magnet faster
- Put more turns on the coil
- Use a stronger magnet

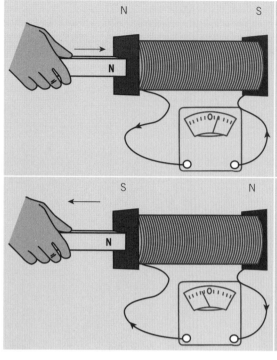

The e.m.f. across the coil can cause a current to flow in the coil if it is in a complete circuit. The current that flows causes the coil to act like a bar magnet. When a north pole is moved towards the coil, this induces a north pole at that end of the coil. **The e.m.f. opposes the change** because the north pole of the coil repels the north pole of the magnet.

When a north pole is moved away from the coil, this induces a south pole at that end of the coil. The south pole of the coil attracts the north pole of the magnet, again **opposing the change**.

Extended

Extended

Relative directions of force, field and induced current

When a force causes a conductor to move in a magnetic field, the e.m.f. induced causes a current to flow if there is a complete circuit. The direction of the induced current is given by **Fleming's right hand rule**.

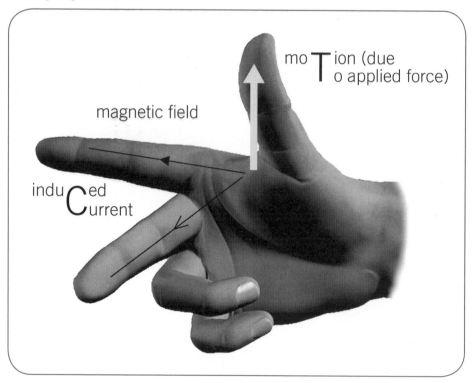

Extended

a.c. generator

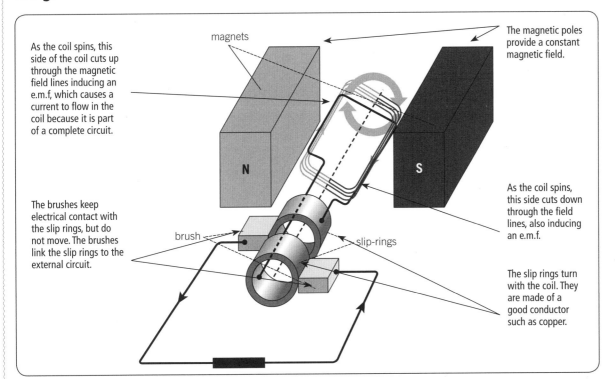

As the coil spins, this side of the coil cuts up through the magnetic field lines inducing an e.m.f, which causes a current to flow in the coil because it is part of a complete circuit.

magnets

The magnetic poles provide a constant magnetic field.

N

S

The brushes keep electrical contact with the slip rings, but do not move. The brushes link the slip rings to the external circuit.

brush

slip-rings

As the coil spins, this side cuts down through the field lines, also inducing an e.m.f.

The slip rings turn with the coil. They are made of a good conductor such as copper.

When the coil reaches the vertical position, the side that was previously cutting up through the field lines will now cut down through the field lines. This means that the **induced e.m.f.** will **change direction** and therefore the **current** will change direction.

Every time the coil reaches the vertical position, the current will change direction. The current produced is **alternating** or **a.c.**.

An alternating current is constantly changing direction, unlike direct current, which flows in one direction around the circuit.

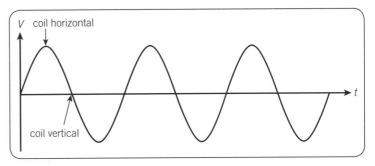

▲ Graph of voltage against time for an a.c. generator

Transformer

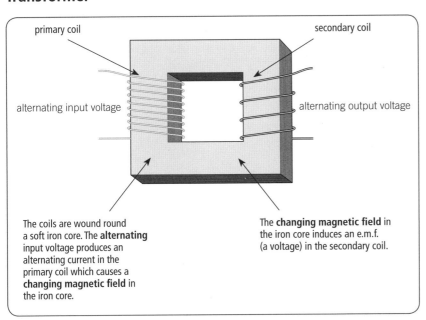

primary coil

secondary coil

alternating input voltage

alternating output voltage

The coils are wound round a soft iron core. The **alternating** input voltage produces an alternating current in the primary coil which causes a **changing magnetic field** in the iron core.

The **changing magnetic field** in the iron core induces an e.m.f. (a voltage) in the secondary coil.

This is a **step-down** transformer. The **output** voltage is **less** than the **input** voltage. This is because there are fewer turns on the secondary coil than on the primary.

A **step-up** transformer has **more** turns on the secondary coil than on the primary. The output voltage is **greater** than the input voltage.

$$\frac{V_s}{V_p} = \frac{N_s}{N_p}$$

V_s = voltage across the secondary coil; N_s = number of turns on the secondary coil (1 turn = 1 loop of wire in the coil);

V_p = voltage across the primary coil; N_p = number of turns on the primary coil

If the transformer is 100 % efficient, the input power is equal to the output power.

$P = IV$ so: $I_pV_p = I_sV_s$

Worked examples

1. The input voltage to a transformer is 20 V. There are 100 turns on the primary coil and 600 turns on the secondary coil. What is the output voltage?

2. The input voltage to a transformer is 300 V and the output voltage is 12 V. There are 100 turns on the secondary coil. How many turns are there on the primary coil?

3. A transformer has 1000 turns on the secondary coil and 200 turns on the primary coil. The output voltage is 10 V. What is the input voltage?

4. The input voltage to a transformer is 50 kV and the output voltage is 100 kV. There are 3 000 turns on the primary coil. How many turns are there on the secondary coil?

Answers

1. $\frac{V_s}{V_p} = \frac{N_s}{N_p}$

   $V_s = V_p \times \frac{N_s}{N_p}$

   $= 20 \times \frac{600}{100}$

   $V_s = 120$ V

   i.e. There are 6 times as many turns on the secondary coil so there is 6 times as much voltage across this coil. Voltage and number of turns are directly proportional.

2. $\frac{V_s}{V_p} = \frac{N_s}{N_p}$

   $N_p = N_s \times \frac{V_p}{V_s}$

   $= 100 \times \frac{300}{12}$

   $N_p = 2500$

   i.e. There is 25 times as much voltage across the primary coil than the secondary and so there are 25 times as many turns on the primary coil

3. $\frac{V_s}{V_p} = \frac{N_s}{N_p}$

   $V_p = V_s \times \frac{N_p}{N_s}$

   $= 10 \times \frac{200}{1000}$

   $V_p = 2$ V

   i.e. There are 5 times fewer coils on the primary then on the secondary coil and so there is $\frac{1}{5}$ of the secondary voltage across the primary

4. $\frac{V_s}{V_p} = \frac{N_s}{N_p}$

   $N_s = N_p \times \frac{V_s}{V_p}$

   $= 3000 \times \frac{100\ 000}{50\ 000}$

   $N_s = 6000$

   i.e. the voltage is stepped up by a factor of 2 and so there are twice as many turns on the secondary coil than the primary.

Extended

Transformers are used to step up the voltage coming from a power station onto the power lines that transmit electrical energy. The power from the power station is constant and so **increasing the voltage decreases the current** (*P = IV* again). Transmitting electrical energy at high voltage and low current reduces the energy lost as heat from the power lines and **increases the efficiency** of the system.

Force on a current–carrying conductor

A wire carrying a current has a magnetic field around it (direction given by the right hand grip rule). If the wire is placed in another magnetic field, the two magnetic fields will interact and there will be a **force** on the wire.

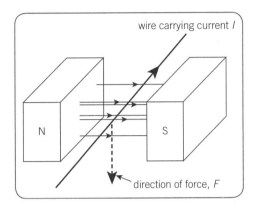

In the above arrangement, the wire will move **downwards** due to the force on it.

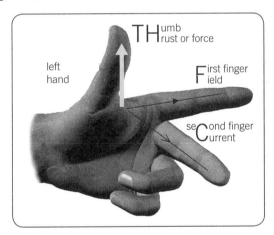

The direction of the force is given by **Fleming's left hand rule**.

To check the direction of the force, hold your fingers in the position shown in the picture. Now turn your hand so that your second finger points into the page and your first finger points from left to right across the page. Your thumb should point down the page, in the direction of the force.

Remember: **F**irst finger **F**ield, se**C**ond finger **C**urrent, thu**M**b for **M**otion

- If the direction of the magnetic field stays the same but the direction of the current is reversed (so that it comes out of the page), the force reverses and is now upwards.
- If the direction of the current remains into the page and the magnetic field is reversed (so that it goes right to left across the page), the force reverses and is now upwards.

Check this example using the left hand rule.

The d.c. motor

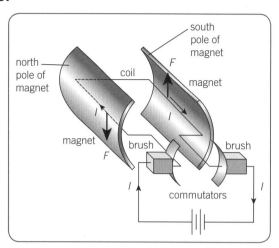

▲ The d.c. motor

Extended

- The poles of the **magnet** are curved to provide a radial magnetic field. This helps to keep the coil in a constant magnetic field.
- There is a force, **F**, on each side of the coil since the coil is carrying a current in a magnetic field. The direction of these forces is given by the left hand rule. The forces cause the coil to **spin**.
- The **commutator** turns as the coil turns. It keeps in electrical contact with the **brushes** and so the current, **I**, keeps flowing through the coil. Every time the coil reaches the vertical position, the two sides of the commutator swap brushes and the **flow of current is reversed**. This means the direction of the **force** on each side of the coil is **reversed** and so the coil keeps spinning in the same direction. Without the commutator, the coil would just oscillate backwards and forwards.

The speed at which the coil spins can be increased by

- increasing the number of turns on the coil
- increasing the current
- increasing the strength of the magnetic field.

Examination style questions

1. a. An experimenter uses a length of wire ABC in an attempt to demonstrate electromagnetic induction. The wire is connected to a sensitive millivoltmeter G.

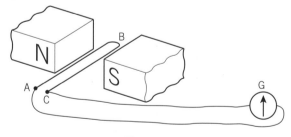

Fig. 1

Using the arrangement in Figure 1, the experimenter finds that she does not obtain the expected deflection on G when she moves the wire ABC down through the magnetic field.

i) Explain why there is no deflection shown on G.
ii) What change should be made in order to observe a deflection on G?

b. Name one device that makes use of electromagnetic induction.

Cambridge IGCSE Physics 0625 Paper 2 Q11 June 2008

2. The diagram below shows a flexible wire hanging between two magnetic poles. The flexible wire is connected to a 12 V d.c. supply that is switched off.

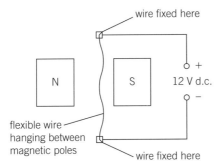

a. Explain why the wire moves when the supply is switched on.

b. State the direction of the deflection of the wire.

c. When the wire first moves, energy transfers from one form to another. State these two forms of energy using an arrow to show the direction of transfer.

d. The diagram below shows the flexible wire made into a rigid rectangular coil and mounted on an axle.

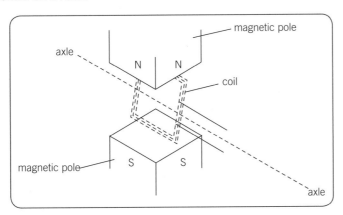

i) Add to a copy of the diagram an arrangement that will allow current to be fed into the coil whilst allowing the coil to turn continuously. Label the parts you have added.

ii) Briefly explain how your arrangement works.

Cambridge IGCSE Physics 0625 Paper 3 Q11 June 2005

3. A transformer is needed to step down a 240 V a.c. supply to a 12 V a.c. output.

a. Draw a labelled diagram of a suitable transformer.

b. Explain

   i) why the transformer only works on a.c.,

   ii) how the input voltage is changed to an output voltage.

c. The output current is 1.5 A. Calculate

   i) the power output

   ii) the energy output in 30 s.

Cambridge IGCSE Physics 0625 Paper 3 Q9 June 2006

Extended

4. a. The transformer in Figure 1 is used to convert 240 V a.c. to 6 V a.c.

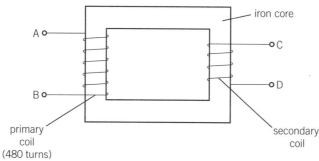

Fig. 1

   i) Using the information above, calculate the number of turns on the secondary coil.

   ii) Describe how the transformer works.

   iii) State one way in which energy is lost from the transformer, and from which part it is lost.

   b. Figure 2 shows a device labelled "IGCSE Transformer".

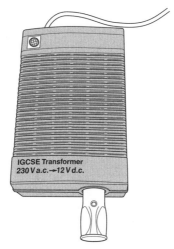

Fig. 2

   Study the label on the case of the IGCSE Transformer.

   i) What is the output of the device?

   ii) From the information on the case, deduce what other electrical component must be included within the case of the IGCSE Transformer, apart from a transformer.

   c. A transformer supplying electrical energy to a factory changes the 11 000 V a.c. supply to 440 V a.c. for use in the factory. The current in the secondary coil is 200 A.

   Calculate the current in the primary coil, assuming no losses from the transformer.

Cambridge IGCSE Physics 0625 Paper 31 Q8 June 2010

5. Your teacher gives you a length of wire, a sensitive millivoltmeter and a powerful magnet. You are asked to demonstrate the induction of an e.m.f. in the wire.
   a. Describe what you would do.
   b. How would you know that an e.m.f. has been induced?
   c. Name a device which makes use of electromagnetic induction.

Cambridge IGCSE Physics 0625 Paper 2 Q10 June 2007

Summary questions on Unit 4

1. What are the units for the following quantities?

Charge	Time period
Current	Frequency
Potential difference or voltage	Energy
Resistance	Power

2. Copy and complete the table to show the names of the electrical components and their symbols.

Switch		Resistor	
	—(V)—		—(A)—
	—⊗—	Thermistor	
Variable resistor		Cell	
	⌐o NO, ⌐o COM, ⌐o NC	Transformer	
Diode			—▥—
		Light dependent resistor (LDR)	

3. Write down the equation linking current, charge and time. Use the correct symbols. Then answer the following questions. Give the answer with the correct unit.
   a. A current of 3 A flows for 6 s. How much charge passes a point in that time?
   b. 6 mC of charge pass a point in a circuit every 100 s. What is the current in the circuit?
   c. How long does it take for 15 C of charge to pass a point when a current of 0.1 A flows?

4. Write down the equation linking resistance, potential difference and current. Use the correct symbols. Then answer the following questions. Give the answer with the correct unit.
   a. What is the potential difference across a 100 Ω resistor when a current of 0.5 A flows through it?
   b. What current flows through a resistor of 330 Ω when a there is a potential difference of 10 V across it?
   c. What value of resistor has a current of 0.8 A flowing through it when the potential difference across it is 1.5 V?

5. Write down the equation linking current, potential difference and power. Use the correct symbols. Then answer the following questions. Give the answer with the correct unit.
   a. What is the power of a lamp when a current of 3 A flow through it and the potential difference across it is 12 V?
   b. A lamp of power 40 W has a potential difference of 230 V across it. What is the current through the lamp?
   c. A lamp of power 100 W has a current of 0.5 A flowing through it. What is the potential difference across the lamp?

6. Explain how an earth wire and a fuse can protect a person from an electric shock.

7. Draw a circuit to vary the potential difference across a lamp and measure the current through the lamp. Include an appropriate component to measure the potential difference across the lamp and a switch to switch the circuit on and off.

8. Draw the symbols for a NOT, AND and OR gate and give their truth tables.

9. Explain how a relay can be used to switch on a large voltage using a small voltage.

10. Draw the electric field lines between charged parallel plates.

11. Copy and complete the paragraph:

When a current carrying conductor is placed in a magnetic _____, there is a _____ on the conductor. The direction of the _____ is given by the _____ _____ rule, where the first finger is the _____, second finger is the _____ and the thumb is for _____. The force on a conductor in a magnetic field can be increased by increasing the _____ through the conductor or increasing the _____ of the _____ _____.

12. Draw and label a diagram of a d.c. motor. Explain how the commutator allows the motor to turn continuously.

13. Describe an experiment to show that the direction of the e.m.f. induced across a conductor depends on the direction that the conductor is moved through a magnetic field.

14. Draw and label a diagram of an a.c. generator. Explain how the slip rings enable the generator to turn continuously.

15. Give three ways in which the output voltage from an a.c. generator can be increased.

16. Draw a diagram of a step up transformer. Label the primary and secondary coils and the iron core. Explain why the transformer will not operate with a d.c. input voltage.

17. Crossword

Across:
1 A component whose resistance decreases with increasing temperature (10)
3 Equal to current × voltage in an electrical circuit (5)
6 An electrical component that limits current (8)
7 Used to determine the direction of the force on a current carrying conductor in a magnetic field (4, 4, 4)
8 Limits electrical current in wiring in the home (7, 7)
11 A property exhibited by the elements iron, cobalt and nickel (14)
12 Produced when a conductor cuts magnetic field lines (7, 3)
15 Repels positive charges (8, 6)
16 Used to step a.c. voltage up and down (11)
17 A quantity whose unit is the ohm (10)
18 Attracts a north pole (5, 4)
19 An electrical component with three terminals (10)

Down:
2 Used to deflect the electron beam in a cathode ray tube (8, 5)
3 A quantity whose unit is the volt (9, 10)
4 Used to switch on a large voltage with a small voltage (5)
5 Used to measure the potential difference across a component (9)
9 Used to vary the potential difference across a component (13)
10 Measures the current through a component (7)
13 A quantity whose unit is the ampere (7)
14 Transforms an input signal of 1 to an output of 0 (3, 4)

Examination style questions on Unit 4

1. Below is a sketch of some apparatus, found in a Science museum, which was once used to show how electrical energy can be converted into kinetic energy.

When the switch is closed the wheel starts to turn.

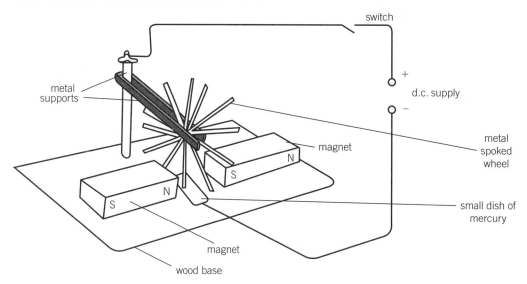

a. Explain why the wheel turns when the switch is closed.
b. On a copy, draw an arrow to show the direction of rotation of the wheel.
c. The d.c. motor is another way to convert electrical energy into kinetic energy. Draw a labelled diagram of a d.c. motor.
d. Describe how the split-ring commutator on an electric motor works.

Cambridge IGCSE Physics 0625 Paper 3 Q9 June 2007

2. The diagram below shows an electrical circuit.

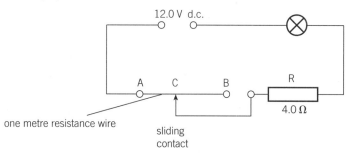

The resistance of the lamp is 4.0 Ω when it is at its normal brightness.

a. The lamp is rated at 6.0 V, 9.0 W.
Calculate the current in the lamp when it is at its normal brightness.
b. The sliding contact C is moved to A. The lamp lights at its normal brightness. Calculate
 i) the total circuit resistance,
 ii) the potential difference across the 4.0 Ω resistor R.
c. The sliding contact C is moved from A to B.
 i) Describe any change that occurs in the brightness of the lamp.
 ii) Explain your answer to (i).
d. The 1 m wire between A and B, as shown in the diagram, has a resistance of 2.0 Ω. Calculate the resistance between A and B when
 i) the 1 m length is replaced by a 2 m length of the same wire,
 ii) the 1 m length is replaced by a 1 m length of a wire of the same material but of only half the cross-sectional area.

Cambridge IGCSE Physics 0625 Paper 3 Q8 June 2006

3. This question refers to quantities and data shown on the circuit diagram.

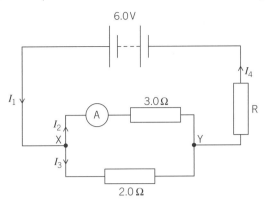

a.   State the relationship between
    i)   the currents I_1, I_2 and I_3,
    ii)   the currents I_1 and I_4.

b.   The ammeter reads 0.80 A. Assume it has zero resistance.

    Calculate

    i)   the potential difference between X and Y,
    ii)   the current I_3,
    iii) the resistance of R.

Cambridge IGCSE Physics 0625 Paper 31 Q9 June 2012

4. a.   The diagram below shows two groups of materials.

    i)   Which group contains metals?
    ii)   Which group contains insulators?
    iii) Write down the name of one of the eight materials above that may be
          charged by rubbing it with a suitable dry cloth.

b.   Two charged metal balls are placed close to a positively charged metal plate.

    One is attracted to the plate and one is repelled.

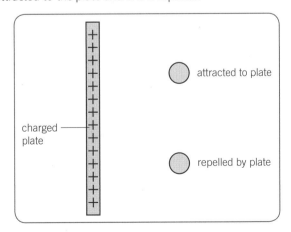

Extended

Copy the diagram and write a + sign on the ball that is positively charged and a − sign on the one that is negatively charged.

c. State what is meant by an "electric field".

Cambridge IGCSE Physics 0625 Paper 2 Q8 November 2006

5. a. Figure 1 illustrates the left hand rule, which helps when describing the force on a current carrying conductor in a magnetic field.

thumb

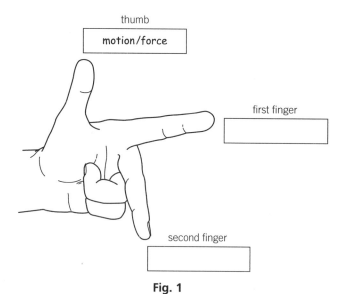

Fig. 1

One direction has been labelled for you.

Copy the diagram and in each of the other two boxes, write the name of the quantity that direction represents.

b. Figure 2 shows a simple d.c. motor connected to a battery and a switch.

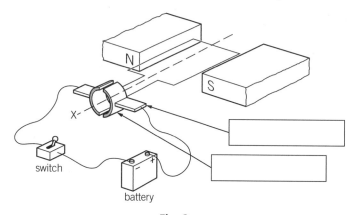

Fig. 2

i) On a copy of the diagram, write in each of the boxes the name of the part of the motor to which the arrow is pointing.

ii) State which way the coil of the motor will rotate when the switch is closed, when viewed from the position X.

iii) State two things which could be done to increase the speed of rotation of the coil.

Cambridge IGCSE Physics 0625 Paper 31 Q9 June 2010

6. a. The diagram below shows two resistors connected to a 6 V battery.

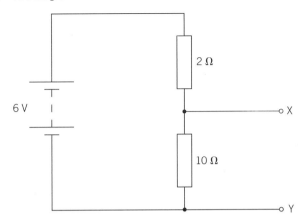

i) What name do we use to describe this way of connecting resistors?

ii) Calculate the combined resistance of the two resistors.

iii) Calculate the current in the circuit.

iv) Use your answer to part **a (iii)** to calculate the potential difference across the 10 Ω resistor.

v) State the potential difference between terminals X and Y.

b. The circuit below is similar to the circuit shown at the beginning of the question, but it uses a resistor AB with a sliding contact.

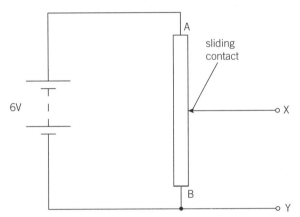

i) State the potential difference between X and Y when the sliding contact is at

1. end A of the resistor,

2. end B of the resistor.

ii) The sliding contact of the resistor AB is moved so that the potential difference between X and Y is 5 V.

On the second diagram, mark with the letter C the position of the sliding contact.

Cambridge IGCSE Physics 0625 Paper 2 Q9 June 2007

5 Atomic physics

5.1 The nuclear atom

KEY IDEAS

✓ The nuclear atom consists of a positively charged nucleus containing protons and neutrons with electrons orbiting
✓ The protons and neutrons are relatively massive compared with the electrons
✓ Protons are positively charged. Neutrons are neutral. Electrons are negatively charged
✓ Isotopes have the same number of protons but different numbers of neutrons

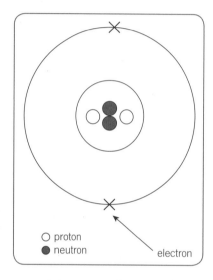

○ proton
● neutron
electron

- The nucleus is in the centre of the atom and contains the **protons** and **neutrons**.
- Protons and neutrons are called **nucleons**.
- The electrons orbit the nucleus.
- The nucleus is **very** small compared to the size of the atom (about 1/10 000 of the diameter).

Relative masses and charges of the particles in an atom:

Name of particle	Relative mass	Relative charge
Proton	1	+1
Neutron	1	0
Electron	1/1800	−1

In a **neutral atom** the number of electrons orbiting the nucleus is equal to the number of protons in the nucleus. Therefore the positive and negative charges cancel out to give **zero** charge.

Number of protons in the nucleus = **proton number** = Z
Number of nucleons in the nucleus = number of protons and neutrons
         = **nucleon number**
         = **A**

A **nuclide** is an atom which is specified by its **proton number** and **nucleon number**.

We represent nuclides like this:

nucleon number $\longrightarrow \; ^{A}_{Z}X \longleftarrow$ symbol for the element

proton number

Example: Carbon-14

Number of protons = 6 = Z
Number of neutrons = 8 = A – Z
Symbol = C

$^{14}_{6}C$

Structure of an atom of carbon-14

This is a diagram representing carbon-14.

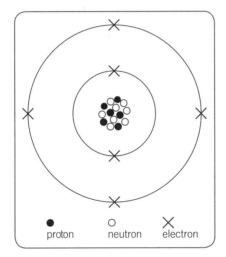

● proton   ○ neutron   ✕ electron

Isotopes

Isotopes of an element have the same number of protons in the nucleus of the atoms, but different numbers of neutrons.

Example: carbon

An atom of the most common isotope of carbon, carbon-12, has 6 neutrons in its nucleus. An atom of carbon-14, shown above, has 8 neutrons in its nucleus. The atom is **neutral**, which means the overall charge on it is zero.

Carbon-12 has a stable nucleus, with the protons and neutrons held together by the strong nuclear force. Having two extra neutrons makes the nucleus of carbon-14 unstable and therefore **radioactive** (see Section 5.2).

The discovery of the nucleus

The nuclear model of the atom was not widely accepted until the early part of the 20th century, when Rutherford and his research team gathered evidence for the existence of the nucleus by carrying out the following experiment.

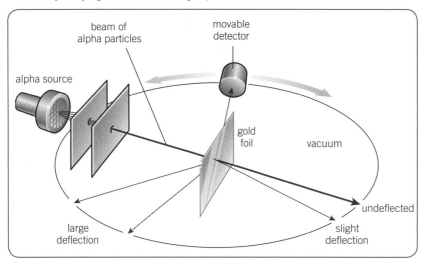

Small positive particles, called alpha particles (see Section 5.2) were fired at a thin gold foil.

Diagram	Description	Conclusion
gold foil beam of alpha particles →	Most of the alpha particles went straight through the gold foil without changing direction.	This showed Rutherford that the atom was mainly space.
gold foil beam of alpha particles →	Some were deflected through small angles	This showed Rutherford that there was positive charge in the atom that repelled the positively charged alpha particles.
gold foil beam of alpha particles →	A few were deflected backwards, towards the alpha source.	The positive charge in the atom must be very concentrated to exert such a large force on the alpha particles. The mass and the positive charge are concentrated at the centre of the atom.

Nuclear fission

In nuclear fission, certain large nuclei split into two smaller nuclei plus some neutrons. In the generation of nuclear power, this process is induced when a neutron is captured by the nucleus, for example:

$$^{1}_{0}n + \,^{235}_{92}U \longrightarrow \,^{143}_{56}Ba + \,^{90}_{36}Kr + 3\,^{1}_{0}n$$

Nuclear fusion

This is the process by which stars (including our Sun) produce heat and light energy. Smaller nuclei are combined to form heavier nuclei, for example:

$$_1^2H + {}_1^3H \longrightarrow {}_2^4He + {}_0^1n$$

Note: $_1^2H$ is deuterium and $_1^3H$ is tritium, both heavy forms of hydrogen.

Examination style questions

1. Uranium-238 can be represented as $_{92}^{238}U$. How many protons, neutrons and electrons are there in an atom of uranium-238?

Adapted from Cambridge IGCSE Physics 0625 Paper 2 Q13 June 2007

2. You have learned about protons, neutrons and electrons.
   a.   How many of each are found in an alpha particle?
   b.   How many of each are found in a beta particle?

Adapted from Cambridge IGCSE Physics 0625 Paper 2 Q12 June 2007

5.2 Radioactivity

A radioisotope is an isotope of an element that has an unstable nucleus and can undergo **radioactive decay** by emitting **alpha** (α), **beta** (β) or **gamma** (γ) radiation (or a combination of the three). By emitting radiation, the nucleus becomes more stable. (For the definition of an isotope see Section 5.1.)

Summary of the properties of alpha, beta and gamma radiation

	α	β	γ
What is it?	helium nucleus (2 protons and 2 neutrons)	fast moving electron	electromagnetic wave (see Section 3.3)
Relative charge	+2	−1	0
Relative mass	4	1/1800	0

Atoms with unstable nuclei occur naturally and emit radiation, contributing to the **background radiation**. Higher doses of radiation can be dangerous because alpha, beta and gamma are all **ionising** and can cause cell damage. This means that when an alpha or beta particle or a gamma ray interacts with an atom, it can remove an electron from the atom. The atom becomes a positive **ion**. Alpha is the most massive and highly charged of the three types of radiation and is by far the most strongly ionising.

Summary of the behaviour of alpha, beta and gamma radiation

	α	β	γ
Ionising effect	strong	weak	very weak
Penetrating effect	Not very penetrating. Absorbed by a few cm of air or a few sheets of paper	More penetrating than alpha. Absorbed by a few mm of aluminium	Very penetrating. Not completely absorbed by lead and thick concrete
Behaviour in electric and magnetic fields	Deflected by electric and magnetic fields	Deflected by electric and magnetic fields, but in the opposite direction to alpha and by a greater amount	Not deflected by electric or magnetic fields

Effect of electric fields

Extended

Alpha particles are attracted to the negative terminal because they are positively charged. Beta particles are deflected in the opposite direction as they are negatively charged. Gamma rays are undeflected.

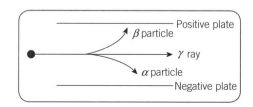

152

Effect of magnetic fields

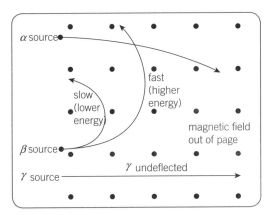

▲ Effect of a magnetic field on alpha, beta and
gamma radiation

Magnetic fields have different effects on alpha, beta and gamma radiation. The deflection
of the alpha particles is given by Fleming's left hand rule (see Section 4.6), with the direction
of motion of the alpha particles as the current. They are deflected at right angles to the
field and to their original direction. The beta particles are negatively charged and so they are
deflected in the opposite direction to the alpha particles. They are deflected more because
they have lower mass. Gamma rays are not charged so they are not deflected at all.

Examination style questions

1. Copy and complete the following table about the particles in an atom.

particle	mass	charge	location
proton	1 unit	+1 unit	In the nucleus
neutron			
electron			

a. Which of the particles in the table make up an alpha particle?
b. Which of the particles in the table is a beta particle?
c. What is the relative mass of an alpha particle?
d. What is the relative charge of an alpha particle?

Cambridge IGCSE Physics 0625 Paper 2 Q12 June 2005

2. The diagram shows a beam of alpha, beta and gamma radiation. The beam passes
between the poles of a very strong magnet.

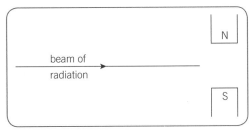

State the direction of deflection, if any, for each type of radiation.

Adapted from Cambridge IGCSE Physics 0625 Paper 3 Q11 June 2006

3. a. Describe what happens when atoms are ionised by ionising radiation.
  b. Explain why alpha radiation is more ionising than beta radiation.

Cambridge IGCSE Physics 0625 Paper 3 Q11 November 2006

Detection of ionising radiation

The Geiger-Müller (G-M) tube works by detecting the ions produced when alpha, beta or gamma radiation enters the tube. It is attached to a counter that registers a count each time a radioactive particle is detected.

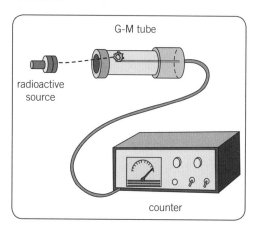

Photographic film is blackened by ionising radiation. The higher the number of radioactive particles incident on the film, the blacker it becomes (see film badge, page 158).

Random nature of radioactive decay

In a sample of radioactive nuclei, there is no way to predict which will be the next nucleus to undergo radioactive decay. Radioactive decay is a **random** process, although there is the same **constant probability** of decay for each nucleus. This is like throwing lots of dice at once. Each of the dice has the same probability of being a six (1/6), but you cannot predict which ones will be sixes each time you throw them.

Nuclear equations

When an unstable nucleus gives out radiation, its composition may change. This can be represented in a **decay equation**. On the left hand side of the equation is the original or **parent nucleus**, on the right hand side is the **daughter nucleus** and the radioactive particle that has been emitted.

Examples

(For an explanation of these symbols, see page 149.)

Alpha decay

$$^{222}_{88}\text{Ra} \longrightarrow \, ^{4}_{2}\text{He} + \, ^{218}_{86}\text{Rn}$$

Note that in alpha decay, the proton number **decreases** by 2 and the nucleon number **decreases** by 4.

Beta decay

$$^{14}_{6}\text{C} \longrightarrow \, ^{14}_{7}\text{N} + \, ^{0}_{-1}\text{e}$$

Note: The nucleon number is unchanged, but the proton number **increases** by 1. This is because in beta decay, a neutron is converted into a proton and an electron. The electron is then fired out of the nucleus, but the proton remains. There is one less neutron and one more proton in the daughter nucleus.

In α and β decay, the nucleus changes to that of a different element.

Gamma decay

$$^{99}_{43}\text{Tc} \longrightarrow \, ^{99}_{43}\text{Tc} + \, ^{0}_{0}\gamma$$

Extended

In gamma decay, the number of neutrons and protons is **unchanged**. The gamma ray takes away some of the excess energy of the nucleus after it has emitted an alpha or beta particle.

Examination style questions

1. The nucleus of sodium-24 decays to magnesium-24 by the emission of one particle. In the equation below the symbol θ represents the emitted particle.

$$^{24}_{11}\text{Na} \longrightarrow {}^{24}_{12}\text{Mg} + {}^{x}_{y}\theta$$

a.   What is the value of x?
b.   What is the value of y?
c.   What is the emitted particle?

Adapted from Cambridge IGCSE Physics 0625 Paper 2 Q12c June 2007

2. The nucleus of uranium-238 decays to thorium-234 by the emission of one particle. In the equation below θ represents the emitted particle.

$$^{238}_{92}\text{U} \longrightarrow {}^{234}_{90}\text{Th} + {}^{x}_{y}\theta$$

a.   What is the value of x?
b.   What is the value of y?
c.   What is the emitted particle?

3. The nucleus of iodine-131 decays to xenon-131 by the emission of one particle. In the equation below θ represents the emitted particle.

$$^{131}_{53}\text{I} \longrightarrow {}^{131}_{54}\text{Xe} + {}^{x}_{y}\theta$$

a.   What is the value of x?
b.   What is the value of y?
c.   What is the emitted particle?

Half-life

As a sample of radioactive material decays, its **activity** decreases with time. The activity is the number of radioactive particles emitted per second. As the number of parent nuclei decreases, the number of radioactive particles emitted per second decreases.

It is not possible to measure the time taken for a sample of radioactive material to completely decay because the activity never falls to zero. Instead, we measure the **half-life**.

The half-life of a radioactive isotope is the time taken for half the nuclei in the sample to decay, or the time taken for the activity of the sample to fall to half of its original value.

Example

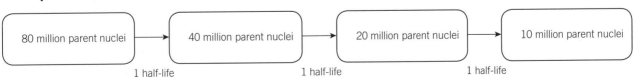

80 million parent nuclei → 40 million parent nuclei → 20 million parent nuclei → 10 million parent nuclei

1 half-life   1 half-life   1 half-life

Each time one half-life passes, the number of parent nuclei halves.

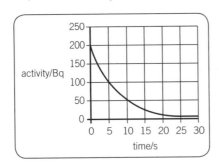

The graph shows how the activity of a sample of radioactive material varies with time. Note that it takes 5 seconds for the activity to halve from 200 to 100 counts per second and then a further 5 seconds to halve from 100 to 50 counts per second. How long does it take for the activity to fall from 50 to 25 counts per second?

Measuring half-life

The half-life of a radioactive isotope (radioisotope) can be measured experimentally, using a Geiger-Müller tube and counter to measure the activity. Before beginning the experiment, several readings of the background activity must be taken (the counts per second before the sample of radioactive isotope is placed in front of the detector). The **background count** must then be **subtracted** from all count rate readings. A graph of corrected count rate (count rate with background count subtracted) on the y-axis versus time on the x-axis is plotted and the half-life found, as for the graph above.

Uses of radioisotopes

The use of a radioisotope depends on the type of radiation it emits and its half-life.

Medical tracers

A **gamma** emitting radioisotope with a **short half-life** (typically 6 hours) is injected into the patient and the gamma radiation that is emitted from his body is measured with a detector called a gamma camera. 'Hot spots' where gamma rays are being emitted at a higher rate show where there is a higher concentration of radioisotope. This allows the medical staff to diagnose a range of conditions, including cancer. Using a radioisotope with a short half-life ensures that the patient does not emit radiation for too long and gamma rays are easily detected outside of the body since they are very penetrating.

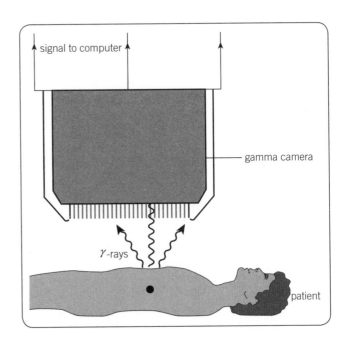

Carbon dating

All living things contain carbon. Many materials, such as cotton and wood, contain carbon because they came from once-living plants. Plants absorb carbon dioxide in the process of photosynthesis and convert it to carbohydrates, which the plant uses to make new plant tissue. However, not all of the carbon that the plant takes in is carbon-12 (the most abundant isotope of carbon). A certain percentage of carbon in the atmosphere is **carbon-14**, a radioactive isotope that emits beta radiation. The half-life of carbon-14 is **several thousand years**. When a tree is alive, it continues to take in carbon so the proportion of carbon-14 remains the same. However, once a tree has been cut down it stops taking in carbon from the atmosphere. The carbon-12 remains, but the carbon-14 decays and its activity drops over a very long period of time. Ancient artefacts containing animal or plant matter such as wood or cloth will have a lower proportion of carbon-14 than when they were made. The activity of the artefact can be measured and compared with what it would have been originally. For example, if the activity has halved, then one half-life has passed and an estimate of its age can be found.

Monitoring thickness

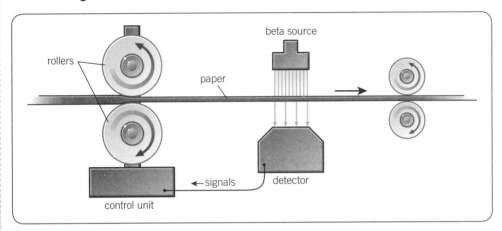

Paper mills use a beta-emitting radioisotope with a long half-life to ensure that the paper remains at the same thickness. The radioactive source is placed at one side of the paper and a detector at the other. Some of the beta radiation is absorbed by the paper but if the paper gets thicker more radiation is absorbed and the count rate at the detector decreases. The rollers adjust automatically to squash the paper and make it thinner again.

Detecting leaks

If there is a suspected leak in a water pipe, a gamma emitter of short half-life (several hours) can be put into the water supply. A detector is held over the area of the suspected leak and the activity measured and compared with background activity. The gamma radiation would be partially absorbed by the metal pipe and only if there was a leak would the reading on the detector be significantly higher than background radiation. After several hours the levels of radiation would be safe again.

Medical therapy

Cobalt-60 is a radioisotope used frequently in radiotherapy. A beam of gamma radiation, emitted from a sample of cobalt-60 is focused with great precision onto a cancerous tumour in the body of the patient. The beam is rotated round the patient to reduce the dose to any one area of the body while the tumour receives a high enough dose to kill the cancer cells.

Safety precautions

▲ Film badge

Ionising radiation can **damage** or **kill cells** or, if changes take place in the DNA in the nucleus of the cell, cause mutations of genes and lead to cancer. Therefore, people who work with radiation must take safety precautions. Radiation workers wear **film badges**, which monitor the dose of radiation received, to ensure that it does not exceed safe levels. Radiotherapists handle radioactive sources in **lead** lined syringes and stand behind a lead screen.

Examination style questions

1. a. Four students attempt to define the *half-life* of a sample of radioactive substance.

   Student A Half-life is half the time for the activity of the sample to decrease to zero.

   Student B Half-life is half the time taken for the activity of the sample to decrease to half its original value.

   Student C Half-life is the time taken for the activity of the sample to decrease to half its original value.

   Which student has given a correct definition?

b. Figure 1 shows two samples of the same radioactive substance. The substance emits β-particles.

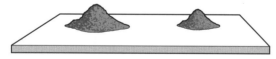

Fig. 1

   Which of the following quantities are the same for both samples?

   the half-life of the samples

   the mass of the samples

   the number of atoms decaying each second

   the number of β-particles emitted per second

c. A quantity of radioactive material has to be taken from a nuclear reactor to a factory some distance away. Figure 2 shows the decay curve for the quantity of radioactive material.

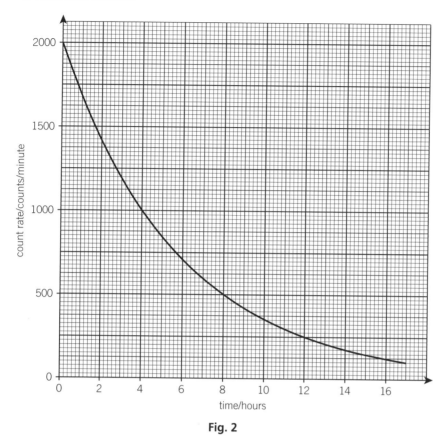

Fig. 2

Just before it leaves the nuclear reactor, the count-rate from the material is 2000 counts/minute. When it arrives at the factory, the count-rate is 1000 counts/minute.

i)   How long did the journey take?

ii)   How many half-lives elapsed during the journey?

iii) The material is only useful to the factory if the activity is at least 100 counts/minute. Use Figure 2 to determine how many hours of useful life the factory has from the radioactive material.

Cambridge IGCSE Physics 0625 Paper 22 Q12 June 2010

2. A student uses a radioisotope that emits only alpha particles and has a long half-life to investigate the percentage of alpha particles that are absorbed by 3 cm of air.

a.   Draw a labelled diagram of the experimental arrangement required to make the determination.

b.   List the readings that the student should take.

Adapted from Cambridge IGCSE Physics 0625 Paper 3 Q10b June 2005

3. A radioactive source, which emits beta-particles, is used as shown in the diagram to detect whether cartons on a conveyor belt have the required volume of pineapple juice in them.

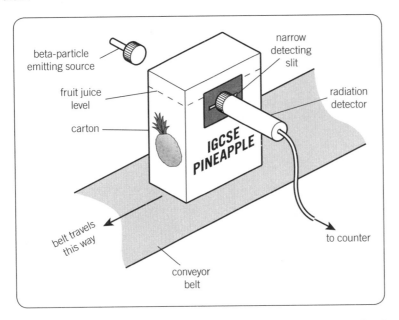

a. State why an alpha-emitting source would not be suitable for this application.

b. State why a gamma-emitting source would not be suitable for this application.

c. The factory has a choice of two beta-emitting sources.

source	half-life
barium-139	85 minutes
strontium-90	28 years

State, giving your reasons, which of these sources is the most suitable for this application.

d. The equipment is set to give a reading of 200 counts/s when there is a carton with the correct amount of pineapple juice between the source and the detector.

Copy the table and tick the appropriate boxes to indicate what reading would be expected in each situation.

	reading		
	more than 200 counts/s	200 counts/s	less than 200 counts/s
carton containing too little juice			
carton containing too much juice			
no carton at all			

Cambridge IGCSE Physics 0625 Paper 22 Q12 June 2011

Summary questions on Unit 5

1. Copy and fill in the blanks

The three types of nuclear radiation are _____, _____ and _____. All are emitted from an unstable _____. The most penetrating type of radiation is _____. The most ionising type of radiation is _____. Alpha radiation cannot penetrate a sheet of _____. Beta radiation is absorbed by a few _____ of aluminium. When entering an electric field, beta particles will be _____. They are attracted to the _____ electric pole. When entering a magnetic field alpha particles are _____ at a _____ angle to their original direction and the direction of the field.

2. Match the statements

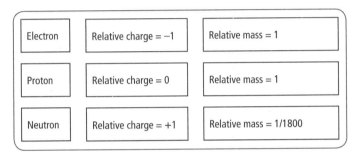

Electron	Relative charge = −1	Relative mass = 1
Proton	Relative charge = 0	Relative mass = 1
Neutron	Relative charge = +1	Relative mass = 1/1800

3. Draw a diagram to show what happened in Rutherford's α scattering experiment. Draw the gold nucleus and a line to indicate the path of the α particles before and after interaction with the gold nucleus.

Most α particles did not get close to a gold nucleus and did not experience a strong repulsive force.

Some α particles came close to a nucleus and were repelled by the positive charge.

Very few α particles hit a gold nucleus head on and were repelled strongly.

4. Copy and complete the table

Isotope	Number of protons in nucleus	Number of neutrons in nucleus
$_{6}^{14}\text{C}$		
$_{2}^{3}\text{He}$		
$_{92}^{238}\text{U}$		
$_{95}^{231}\text{Am}$		

5. Crossword

Across:
4 The part of the atom from which radioactive particles are emitted (7)
6 Alpha radiation is absorbed by this (5)
8 This must be taken before an experiment into radioactive decay is carried out (10, 5)
9 The name of the model of the atom that came before the nuclear model (4, 7)
12 An isotope of this element is used as 'fuel' in a nuclear power station (7)
14 A helium nucleus (5, 8)
16 Necessary to absorb gamma radiation (4)
18 This disease can be treated with radiation (6)

Down:
1 Knock an electron off an atom (6)
2 What an atom is mainly made up of (5)
3 The splitting of uranium nuclei into smaller nuclei and neutrons (7)
5 The source of background radiation that becomes more dangerous at high altitiude (6, 4)
7 This detects radioactive particles in a film badge (12, 4)
8 A fast moving electron (4, 8)
10 An electromagnetic wave (5, 3)
11 An instrument used to detect radioactive particles (6, 7)
13 A physicist who suggested experiments to see what would happen when alpha particles were fired at a gold foil (10)
15 The time taken for the activity of a sample of radioisotope to halve (4, 4)
17 An isotope of this element is used for dating artifacts such as ancient clothing (6)

6. Draw a mind map (spider diagram), including all the important points from Unit 5. Use diagrams, colour coding and mnemonics to help you remember the key points. Ensure that you group the key ideas logically. When you have finished, ask someone to test you on the content of your mind map.

Examination style questions on Unit 5

1. A beam of ionising radiation, containing α-particles, β-particles and γ-rays, is travelling left to right across the page. A magnetic field acts perpendicularly into the page.

a. In the table below, tick the boxes that describe the deflection of each of the types of radiation as it passes through the magnetic field. One line has been completed, to help you.

	not deflected	deflected towards top of page	deflected towards bottom of page	large deflection	small deflection
α-particles		✓			✓
β-particles					
γ-rays					

b. An electric field is now applied, in the same region as the magnetic field and at the same time as the magnetic field.

What is the direction of the electric field in order to cancel out the deflection of the α-particles?

Cambridge IGCSE Physics 0625 Paper 31 Q11 June 2009

2. a. The graph below is the decay curve for a radioactive isotope that emits only β-particles.

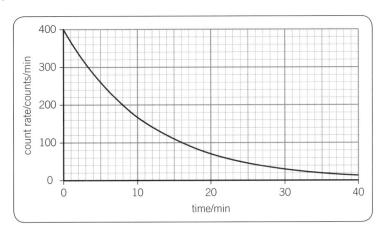

Use the graph to find the value of the half-life of the isotope.

Indicate, on the graph, how you arrived at your value.

b. A student determines the percentage of β-particles absorbed by a thick aluminium sheet. He uses a source that is emitting only β-particles and that has a long half-life.
   i) Draw a labelled diagram of the apparatus required, set up to make the determination.
   ii) List the readings that the student needs to take.

Cambridge IGCSE Physics 0625 Paper 3 Q10 June 2005

3. The table below contains some information about uranium-238.

proton number $Z = 92$

nucleon number $A = 238$

decays by emitting α-particle

a. State how many electrons there are in a neutral atom of uranium-238.
b. State where in the atom the electrons are to be found.
c. State how many neutrons there are in an atom of uranium-238.
d. State where in the atom the neutrons are to be found.
e. State what happens to the number of protons in an atom of uranium-238 when an α-particle is emitted.

Cambridge IGCSE Physics 0625 Paper 2 Q11 November 2006

4. a. State what is meant by
    i) the "half-life" of a radioactive substance,
    ii) "background radiation".
b. In a certain laboratory, the background radiation level is 25 counts/minute.
Below is a graph of the count-rate measured by a detector placed a short distance from a radioactive source in the laboratory.

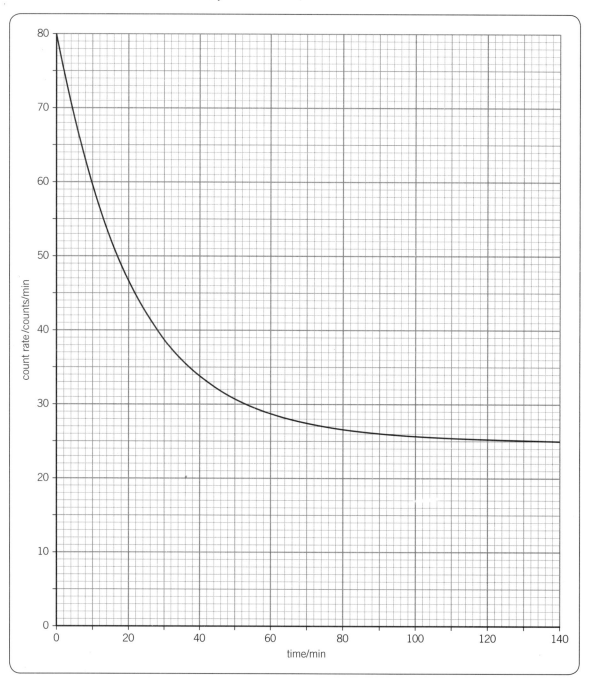

i)   At zero time, the measured count-rate of the source and background together
     is 80 counts/minute.
     Calculate the count-rate due to the source alone.
ii) After one half-life has elapsed, what is the count-rate
     1. due to the source alone,
     2. measured by the detector?
iii) Use the graph to find the half-life of the source.
iv) Why does the graph **not** drop below the 25 counts/minute line?
v)   Sketch the curve that might be obtained for a source with a shorter half-life.

Cambridge IGCSE Physics 0625 Paper 2 Q12 November 2005

Answers to questions

1 General physics

Page 8

1. Check that the stopwatch is reading zero; start stopwatch as the piston moves through the centre; count at least 20 complete cycles before stopping stopwatch; divide time by number of complete cycles.

Page 13

1. a Between t = 30 and t = 45 mins
   b 7.5 mins
   c i distance travelled = area under graph between
      $t = 0$ and $t = 12\frac{1}{2}$ mins
   ii average speed = total distance travelled divided by the total time for journey divided by 60 mins

2. a $\Delta v = 32$ m/s; a =10 m/s^2;
   $t = \frac{\Delta v}{a} = \frac{32}{10} = 3.2$ s
   b
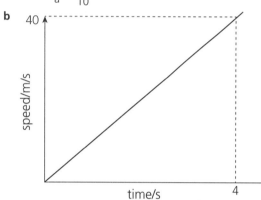

Page 14

3. a OP – constant acceleration; PQ – constant acceleration (greater than OP); QR – constant speed; RS – constant deceleration;
   b O and S
   c 6 m/s
   d 70 s
   e Total distance travelled = area under graph
   f $a = \frac{2}{50} = 0.04$ m/s^2

4. a $u = \frac{d}{t} = \frac{25}{2} = 12.5$ m/s
   b speed is decreasing
   c time for tree 3 to tree 4 is greater than for time for tree 2 to tree 3

Page 15

5. a Stage 2: $t = \frac{4800}{12} = 400$ s
   b
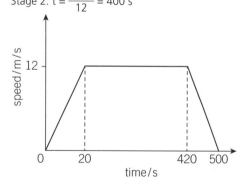

---

c Total distance travelled:
$$= (\tfrac{1}{2} \times 20 \times 12) + (400 \times 12) + (\tfrac{1}{2} \times 80 \times 12)$$
$$= 120 + 4800 + 480 = 5400\,\text{m}$$

d average speed = $\frac{\text{total distance}}{\text{time taken}} = \frac{5400}{500} = 10.8$ m/s

e $a = \frac{\Delta v}{t} = \frac{12}{20} = 0.6$ m/s^2

Page 18

1. Weigh the car using the Newton meter;
   find the mass from $mass = \frac{weight}{10}$ m/s^2
   (for rest of question see description on p.17)

2. (For first part see description on p.17)
   To find the density of cork the mass, M, could still be found using the balance, but the volume must be calculated by finding its dimensions and using *cross-sectional area × length* to calculate the volume.
   Area = area of a circle, πr^2;
   d = 2 × radius = 2r;
   Volume = $\pi r^2 l$;
   Density = $\pi r^2 l / m$;

3. a Volume = Reading Q – Reading P
   b Mass = Reading S – Reading R
   c i $\rho = \frac{m}{v}$
   ii $p = \frac{57.5}{25} = 2.3$ g/cm^3

Page 19

Practical question

1. l = 3.6 cm

2. $A = \frac{\pi\,(0.6)^2}{4} = 0.28$ cm^2

3. $V = \pi r^2 l = 1.0$ cm^3
   Volume of bundle = 10 × 1.0 = 10.0 cm^3
   $P = \frac{m}{v} = \frac{59.1}{10.0} = 5.91$ g/cm^3

Page 23

1. a length = 8.5 cm
   b length = 19 cm
   c extension = length – initial length = 19 – 8.5 = 10.5 cm

2. a They have both size and direction
   b f = m × a = 6 N

3. a Q = elastic limit
   b The extension is directly proportional to the force applied
   c In the region QR, the extension is greater for the same increase in applied force
   d $k = \frac{F}{x}$ = gradient of graph = $\frac{8}{2}$ = 4 N/m

Page 24

4. a i $a = \frac{28.5}{3} = 9.5$ m/s^2
   ii $d = \frac{1}{2} \times 3 \times 28.5 = 42.75$ m
   iii terminal velocity = 15 m/s

b Plastic ball larger so drag is greater.
For the rubber ball, this force not big enough to balance weight.
For the plastic ball, drag is big enough to balance weight.

c W = mg = 0.05 × 10 = 0.5 N

Page 25

5. c extension = 67 − 40 = 27 cm

d load = 2.5 N; weight = $\frac{load}{g}$ = 0.25 kg = 250 g

Page 26

Practical question

1. m/g; θ/°

2. Graph is a straight line through (0,0)

Page 28

1. a The magnitude of the force and the perpendicular distance of the force from the pivot

b i 1: force; 2: moment

ii $F = F_1 + W + F_2$

iii F

2. a Resultant force = 0; resultant moment = 0

b 6 × 40 = F × 30

$F = \frac{240}{30}$ = 8.0 N

c Force = 0.5 N; downwards

Page 29

Practical question

1. Use a set square

2.

F/N	d/m	$\frac{1}{d}$/$\frac{1}{m}$
0.74	0.900	1.11
0.78	0.850	1.18
0.81	0.800	1.25
0.86	0.750	1.33
0.92	0.700	1.43

3. c G = 0.53 N m

4. W = $\frac{0.53}{0.490}$ = 1.1 N

Page 31

1. a Line of action of weight of box acts outside its base

b i less than

ii the centre of mass is higher and so the line of action of the weight will act outside the base at a lower angle

Page 33

1. a ii Force = 5600 N plus or minus 100 N angle = 30° plus or minus 2°

b i it has direction (as well as magnitude)

ii e.g velocity, acceleration, momentum, displacement

Page 39

1. a The energy supplied is transformed into heat, light and sound

b i 24 kJ

ii Dissipated as heat into the surroundings

iii 696 kJ

iv efficiency = $\frac{24}{720}$ = 3% ie very low

2. a i gravitational potential

ii 1: weight of basket; 2: height through which it is lifted

b chemical

c time taken to lift the basket

Page 42

1. a gravitational potential

b kinetic

c kinetic

d electrical

e heat

Page 44

1. a WD = f × d = mgh

b i WD = 100 × 8 = 800 J

ii power = $\frac{WD}{t}$ = $\frac{800}{5}$ = 160 W

iii Increasing internal energy

Page 45

2. a i WD = mgh = 50 × 10 × 4 = 2000 J

ii P = $\frac{WD}{t}$ = $\frac{2000}{20}$ = 100 W

b gravitational potential → kinetic → heat

Page 48

1. a Skate, since area is smaller for the same weight so pressure is greater.

b The area of the vehicle is large and so for a constant wind pressure, the force is high and could blow the vehicle over

2. a i p = $\frac{F}{A}$ = $\frac{84}{6.0 \times 10^{-5}}$ = 1.4 × 10⁶ Pa

ii 84 N

iii area is smaller therefore pin exerts far greater pressure

b i p = ρgh = 1000 × 10 × 3 = 30 000 Pa

ii 30 000 Pa

Pages 49–50

Summary questions on Unit 1

1. distance; time; m/s; gradient; direction; magnitude; rate of change; m/s²; gradient; distance travelled

2. 2 m/s

3. 3 m/s²

4. speed

5. distance travelled

6. Weight is a force that acts on a mass in a gravitational field

7. 4 g/cm³

8. shape, directly proportional, proportionality, accelerate, doubled, resultant, subtracted, added, direction, speed, perpendicular, circle

9. 5 cm

10. kg, force, newtons, mass, gravitational field strength, volume, g/cm³, kg/m³

11. kettle – electrical to thermal
generator – kinetic to electrical
battery – chemical to electrical
solar cell – light to electrical

12. 4.9 J

13. 2500 J

14. 5 m

15. 200 W

16. 36000 N/m²

17. 0.6 N/cm²

Page 51

Examination style questions on Unit 1

1. a

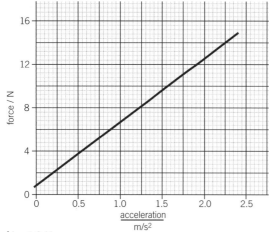

b i 1.2 N

ii trolley does not move

c $\text{gradient} = \dfrac{\text{y step}}{\text{x step}} = \dfrac{(14-1.2)}{2.25} = 5.69$

d i $F = ma$

ii $m = \dfrac{F}{a} = 5.7 \text{ kg}$

e

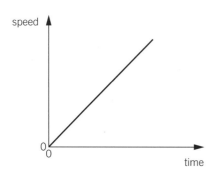

Page 52

2. a 1.81 s or 1.8 s, mean value

b Time a minimum of 2 oscillations; divide result by the number of oscillations B1; repeat at least three times and find mean; time with reference to fixed / fiducial point or top or bottom of oscillation; zero the stop-watch.

3. a

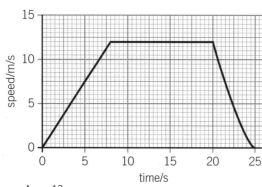

b $a = \dfrac{\Delta v}{t} = \dfrac{12}{8} = 1.5 \text{ m/s}^2$

c $\text{distance} = \dfrac{1}{2} \times 5 \times 12 = 30 \text{ m}$

d $f = mxa = 4000 \times 1.2 = 4800 \text{ N}$

e For the same accelerating force, there is a greater mass due to the extra passengers, and so acceleration is less.

Page 53

4. a i

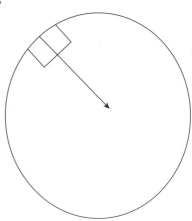

ii Force increases

b i

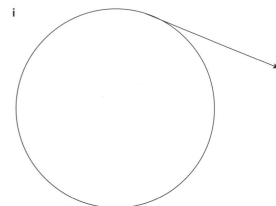

ii The force of friction was not large enough to provide the centripetal force.

c i steady speed

ii $\text{circumference} = 2\pi r = \pi d = 20.4 \text{ cm}$

iii $\text{increase in speed per second} = \dfrac{25}{3} = 8.3 \text{ m/s}^2$

Page 54

5. a i gradient of speed-time graph is constant

ii $a = \dfrac{\Delta v}{t} = \dfrac{6}{8} = 0.75 \text{ m/s}^2$

b i decreases

ii friction force is equal and opposite to gravitational force

iii

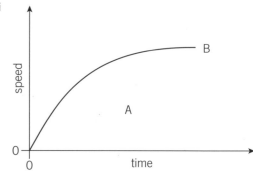

Page 58

1. **a** Constant random motion
   b As the temperature increases, the velocity of the particles increases; particles collide with walls of container more often and with greater force and so the pressure increases.
2. **a** Evaporation
   b Particles with greatest KE escape from surface of water, reducing the volume of water.
3. **a** volume of rubber cover decreases, so pressure increases
   b $1 \times 60 = 1.5 \times V$
      $V = 40 \text{ cm}^3$
      So the reduction in volume = 20 cm³
   c average speed of mols/particles/atoms greater at high temp and so there are more collisions/collisions with greater force; pressure increases.

Page 59

4. **a** The dust particle is constantly bombarded by the air molecules, which are in constant random motion.
   b **i** The ones with the greatest kinetic energy.
      ii These have the energy required to overcome the attractive forces between molecules.
5. **a** **i** Constant, random motion
      ii Collides with container walls putting force on walls.
   b **i** Increases
      ii Remains constant
   c **i** Movement with greater KE in gas.
      ii Separation of molecules in gas is greater.

Page 60

Practical question

1. The temperature of thermometer 2 decreases more rapidly than that of thermometer 1, due to the evaporation of the water.
2. The temperature of thermometer 3 decreases more rapidly than that of thermometer 2, because acetone evaporates more rapidly than water.

Page 63

1. **a** Choose a different liquid, one with a larger range of temperature between its freezing and boiling point.
   b Increase the number of scale divisions and have a thinner thread in the container.
2. **a** The piston must move to the right.
   b The heat supplied to the piston is used to do work in pushing the piston to the right; the temperature remains constant.

Page 64

3. **a** They vibrate with greater kinetic energy and move further apart.
   b **i** Thermometers
      ii Railway tracks

Page 69

1. **a** Mass of aluminium on balance; change in temperature with thermometer; energy delivered to the aluminium.
   b Higher.
   c Some heat is lost to the surroundings.

Page 70

2. **a** Evaporation occurs over a range of temperatures; boiling occurs at a specific temperature (the boiling point).
      Evaporation decreases the temperature of the liquid; the temperature of the liquid remains constant during boiling.
   b Energy is required to break the bonds between the liquid particles; all the heat energy is used to do work against the surroundings, not to increase the internal energy of the liquid.
   c $E = P \times t = 100 \times 20 \times 60 = 120000 \text{ J}$
      $120000 = \Delta m \times L = 0.05 \text{ L}$
      $L = 2.4 \times 10^6 = 2400 \text{ kJ/kg}$
3. **a** **i** 20°C, 15°C
      ii As the temperature of the water increases, more heat is lost to the surroundings and so the temperature rises at a lower rate.
   b $c = \dfrac{E}{(m \times \Delta T)} = \dfrac{60 \times 210}{(0.075 \times 40)} = 4200 \text{ J/(kg °C)}$
   c

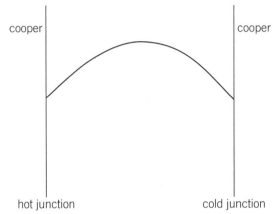

4. **a** **i** freezing, solidification, condensation e.g. water to ice, steam to water, gas to solid
      ii No change
   b Energy required to change temperature of the body by 1 °C / 1 K or mass × specific heat capacity
   c **i** $E = m \times c \times \Delta T = 250 \times 4.2 \times 20 = 21000 \text{ J}$
      ii 21000 J
      iii $E = \Delta m \times L$
         $21000 = m \times 330$
         $m = \dfrac{21000}{330} = 63.6 \text{ g}$

Page 71

Practical question

1. 24 °C
2. Beaker A: 8 °C; beaker B: 5 °C
3. The Bunsen burner loses more heat to the surroundings; the power of the fixed voltage source is greater than the Bunsen burner.

Page 75

1. Take the starting temperature on the thermometers; read the thermometers after a fixed interval of time; the best absorber of heat radiation will have a higher temperature.

Page 76

2. **a** Radiation
   b Conduction

Page 78

1. **a** **i** convection
   ii hot water expands and the molecules move further apart
   hot water is less dense and rises
   cool water is more dense and sinks
   b hot air rises
2. **a** **i** Copper is a good conductor of heat.
   ii Plastic is an insulator.
   b **i** Conduction
   ii Radiation and convection.
   c equal to 40 W

Page 79

3. **a** **i** Conduction
   ii The particles at A vibrate faster and pass on the increased vibrations to neighbouring particles.
   b 4 different coloured surfaces are heated at equal distances from a radiant heater (see diagram page 75); the initial and final temperatures of each of the surfaces are taken after equal time intervals.

Page 80

Practical question

1. t/s, θ/°C
2. Place insulation on the bottom of the beaker; cover the top of the beakers.

Pages 81–82

Summary questions on Unit 2

1. Conduction – transfer of vibrational energy from particle to particle; convection – movement of hot fluid due to changes in density; radiation – an electromagnetic wave; evaporation – most energetic particles escape from the surface of a liquid.
2. melt, 1, boil, 1, temperature, 1
3.

	Solid	Liquid	Gas
Structure	Structure – see diagrams page 55.		
Arrangement of molecules	Close together; vibrating in fixed ositions	Close together but able to move past one another	Far apart; in constant random motion
Compressible?	No	No	Yes
Flows?	No	Yes	Yes
Shape	Keeps its shape	Fills the bottom of the container	Fills the whole container

4. Arrows point clockwise.
5. E = m × c × ΔT
   $c = \dfrac{E}{(m \times \Delta T)} = \dfrac{4500}{0.5 \times 10} = 900$ J/(kg °C)
6. E = Δm × L
   $L = \dfrac{66\,000}{0.2} = 330000$ J/kg
7. The pressure increases since the volume decreases at constant temperature. Collisions between the particles and walls of the container occur more frequently.
8. Across: radiation, specific heat capacity, kelvin, conduction, free electrons, gas, thermocouple, melting point, absolute zero
   Down: inversely proportional, evaporation, alcohol, convection, kinetic, boiling point, matt black

Page 83

Examination style questions on Unit 2

1. **a** Reading 2 = mass of beaker + stirrer + thermometer + water
   Reading 3 = mass of beaker + stirrer + thermometer + water + melted ice
   b **i** heat lost = mass of water × 20 × specific heat capacity of water
   ii heat gained = mass of ice × specific latent heat of fusion
   c $L = \dfrac{12800}{0.030} = 4.3 \times 105 = 430000$ J/kg
   d The ice may not be at 0°C; heat may be taken in from the surroundings to melt the ice.
2. **a** **i** smaller because area smaller
   ii smaller because depth/height smaller
   b **i** p = hρg = 12 × 1000 × 10 = 1.2 × 10⁵ Pa
   ii 1.2 × 10⁵ Pa + 1.0 × 10⁵ Pa =2.2 × 10⁵ Pa
   iii 1.1 cm³
   iv volume will be bigger

Page 84

3. **a** More energetic molecules escape from surface and molecules left have lower average energy so temperature is lower or heat needed to evaporate comes from remaining liquid
   b **i** Dull surface is a better radiator
   ii C hotter so it evaporates at a faster rate in C
   iii Less liquid in D
   iv E has a greater surface area and so greater rate of loss of heat by evaporation/convection/ conduction/radiation

Page 89

1. **a** **i** decreasing
   ii amplitude decreasing
   b **i** constant
   ii waves equally spaced
   c **i** 13 ± 1
   ii 1. 300 waves passing a point per second;
   2. $\dfrac{1}{300} = 0.0033$ s; 3. $\dfrac{1}{300} \times 13 = 0.04$ s
   d **i** yes
   ii yes
   iii no
2. **a**

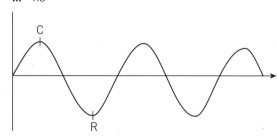

   b Air molecules are vibrating, alternately moving closer together and then further apart.
   c $\lambda = \dfrac{v}{f} = \dfrac{330}{500} = 0.66$ m

Page 93

1. **a** Angle of incidence – 0° i.e. along normal
   b Speed decreases, frequency is constant, wavelength decreases.

c

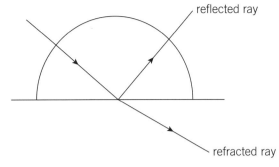

reflected ray

refracted ray

d $\dfrac{\sin 30°}{\sin r} = 0.67$

$r = \sin^{-1}\left(\dfrac{\sin 30°}{0.67}\right) = 48°$

Page 94

2. a i Diagram to show boundary, normal and ray
bending towards normal
Angle of incidence labelled i or 51°
Angle of refraction labelled r or 29°

ii $n = \dfrac{\sin i}{\sin r} = \dfrac{\sin 51}{\sin 29}$

$n = 1.603$

b Ray is totally internally reflected
Angle of incidence is greater than the critical angle
Critical angle = 38.6°

3. a

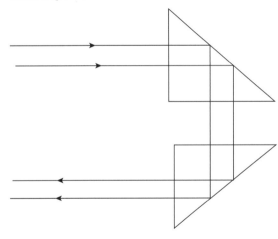

b Virtual, inverted, same size

c Angle of incidence = 0°

d speed $= \dfrac{3 \times 10^8}{1.5} = 2 \times 10^8$ m/s

e Angle of incidence at b> critical angle

Page 97

1. a

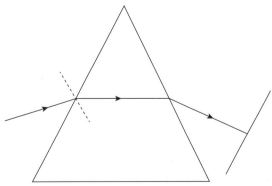

b i dispersion
ii red
iii violet

Page 98

2.

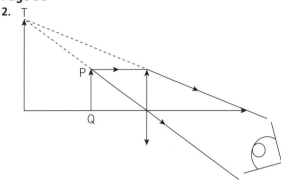

3. a See diagram for question 2

b virtual upright/erect magnified/enlarged further
from lens

Page 100

1. a i gamma or x-ray
ii infrared or radio

b $f = \dfrac{v}{\lambda} = \dfrac{3 \times 10^8}{1.0 \times 10^{-14}} = 3.0 \times 10^{22}$ Hz

c 3×10^8 m/s

Page 102

1. a longitudinal movement
b 8.7–8.9 cm
c more waves in same distance/shorter wavelength

2. a $\dfrac{2000}{60} = 33$ times per second

b Frequency = 33 Hz

c Yes, hearing range is typically 20 Hz – 20 kHz

Page 103

3. a i A building/wall
ii $d = u \times t = 320 \times 1.5 = 480$ m

b $v = \dfrac{d}{t} = \dfrac{6}{5} = 1.2$ m/s

Pages 104–106

Summary questions on Unit 3

1. longitudinal; transverse; transverse; longitudinal; vibrations;
parallel; wavelength; time period; waves; second; v = fλ;
reflection; refraction; diffraction.

2.

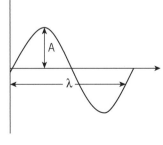

3.

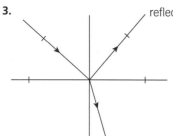 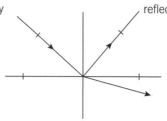 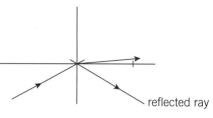

reflected ray

reflected ray

reflected ray

4. **a** $\sin r = \sin\dfrac{56}{1.5}$

$r = 34°$

angle of reflection = 56°

b $\sin r = \sin\dfrac{35}{0.67}$

$r = 59°$

angle of reflection = 35°

c angle of reflection = 60°

There is no refracted ray.

5. Ratio of speed of light in air to speed in material – refractive index; wave spreading out as it passes through a gap – diffraction; wave changing direction as its speed changes – refraction; light bouncing off a smooth surface – reflection; white light splitting into a spectrum as it is deviated by a prism – diffraction.

7. **a** $\lambda = \dfrac{v}{f} = \dfrac{3 \times 10^8}{1 \times 10^{16}} = 3 \times 10^{-8}$ m UV

b $f = \dfrac{v}{\lambda} = \dfrac{3 \times 10^8}{0.1} = 3 \times 10^9$ Hz radio

c $\lambda = \dfrac{v}{f} = \dfrac{3 \times 10^8}{1 \times 10^{20}} = 3 \times 10^{-12}$ m gamma

8. Across: gamma, violet, ultra-violet, normal, amplitude, critical angle, infra-red, sound, radio, frequency, wavelength
Down: refract, microwave, focal length, virtual, optical fibre, hertz, rarefaction, compression, reflect, convex

9. Different wavelengths of light refract by different amounts in a denser medium, causing white light to split into a spectrum.

10.

11. $f = \dfrac{v}{\lambda} = \dfrac{330}{0.1} = 3300$ Hz

speed = $3300 \times 0.45 = 1485$ m/s

12. $u = \dfrac{d}{t} = \dfrac{2}{0.00042} = 4760$ m/s

Page 106

Examination style questions on Unit 3

1. **a** **i** hits surface at right angles so the angle of incidence = 0

ii reflection shown at second surface at 45°

b **ii** $i = r$ in symbols or words NOT $\sin i = \sin r$

Page 107

2. **a** $t = \dfrac{d}{u} = \dfrac{498}{332} = 1.5$ s

b 0.75 s, 2.25 s

3. **b** **i** $\dfrac{3 \times 10^8}{\text{speed in glass}} = 1.5$; speed in glass = 2×10^8 m/s

ii $\dfrac{\sin 70°}{\sin r} = 1.5$

$r = 38.8°$

Page 108

4. **a** **i**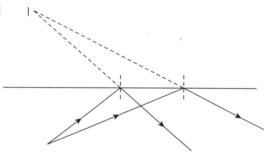

ii same size, upright, virtual, laterally inverted

5. **a** Stop clock, starting pistol

b Start stopclock when smoke from pistol is seen; stop stopclock when sound is heard; measure distance between student with starting pistol and observer; time of travel is read on the stop clock.

c speed = distance / time

d Repeat the experiment several times, eliminate anomalous results and average

e **i** 100 m/s

ii 1000 m/s

6. **a**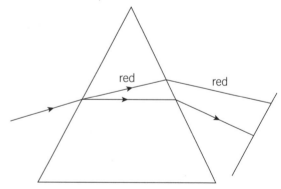

red red

b $\sin r = \dfrac{\sin 40°}{1.52} = 25°$

c **i** 3×10^8 m/s

ii 3×10^8 m/s

Page 111

1. **a** **i** iron rod
   ii plastic rod
   b S S N
   c arrow vertically up in each case

Page 112

2. **a** **i** iron
   ii 1: nothing; 2: nothing

   b

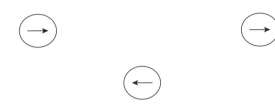

3.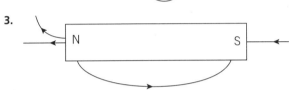

Page 117

1. **a** 6.3 V; 330 Ω
   c **iii** A best fit line allows values of time or resistance between the plotted values to be found with greater accuracy.
   d Temperature is greatest for the lowest value of resistance on the graph.

Page 118

2. **a**

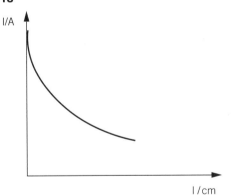

   b $I = \frac{V}{R} = \frac{2.0}{5.0} = 0.4$ A
   c **i** 20 Ω
      ii $I = \frac{V}{R} = \frac{2.0}{20.0} = 0.1$ A
   d 5.0 Ω
   e heating, magnetism

Page 119

3. **a**

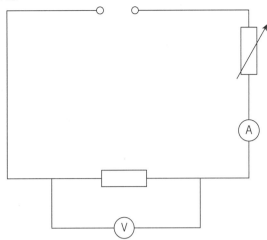

   b Vary resistance of variable resistor to obtain 5 different voltage readings, taken on voltmeter. Take 5 corresponding values of current on the ammeter.
   c **i** $R = \frac{V}{I} = \frac{6}{0.5} = 12\,\Omega$ unknown resistor = 9 Ω
      ii $Q = I \times t = 0.5 \times 120 = 60\,C$
      iii $P = I \times V = 0.5 \times 1.5 = 0.75\,W$

4. **a** **i** electrons
      ii negatives on right, positives on left
      iii remove insulating base/add metal wire from S to earth so that negatively charged electrons flow from S to earth
   b **i** $\frac{30}{(2 \times 60)} = 0.25A$
      ii $E = QV = 30 \times 1.5 \times 10^6 = 4.5 \times 10^7$ J

Page 120

Practical question

1. AB = 50 cm; AC = 75 cm; AD = 100 cm;
3. AB: $R = \frac{V}{I} = 2.53\,\Omega$

   AC = 4.00 Ω
   AD = 5.20 Ω
4. 2.5 Ω, 4.0 Ω, 5.2 Ω
5. R / Ω
6. $R = 5.2 \times 1.5 = 7.8\,\Omega$

Page 126

1. **a** **i** meter 2
      ii ammeter
   b **i** meter 1
      ii voltmeter
   c **i** 1.6 V
      ii $R = \frac{V}{I} = \frac{1.6}{0.8} = 2.0\,\Omega$
      iii straight line through origin
      iv greater slope
      v wire B. Larger resistance from longer wires

Page 127

2. **a i** current

   ii p.d.

   b R = R1 + R2

   $I = \dfrac{9.0}{4.8} = 1.875$ mA or $\dfrac{9.0}{4800} = 1.875 \times 10^{-3}$ A

   Voltmeter reading = 6.75 V

   c temperature of thermistor rises and its resistance falls; current increases; magnetic field of relay closes switch (and bell rings)

3. **a i** $R = \dfrac{V}{I} = \dfrac{1.8}{0.45} = 4.0\ \Omega$

   ii $E = IVt = 0.45 \times 1.8 \times 9 \times 60 = 437.4$ J

   b R second coil $= \dfrac{1.8}{0.3} = 6\ \Omega$ ie. increased by factor 1.5 diameter $\times$ ½, increases R by factor 4

   Length of second coil is $\dfrac{1.5}{4} = 0.375$ times length of first coil

Page 128

4. **a**

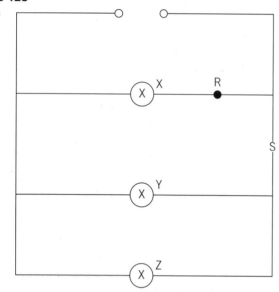

   d $R = \dfrac{V}{I} = \dfrac{12}{3} = 4\Omega$

   e i Parallel circuit

   ii 4.0 A

5. **a i** 0 A

   ii 12 V

   b i $I = \dfrac{V}{R} = 0.5$ A

   ii $8 \times 0.5 = 4$ V

   c $\dfrac{1}{R1} + \dfrac{1}{R2} = \dfrac{1}{R}$

   R = 5.3 Ω

   $I = \dfrac{12}{5.3} = 2.25$ A

Page 129

6. **a** Region around an electrostatic charge where another charge experiences a force.

b + −

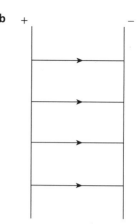

c $I = \dfrac{Q}{t} = \dfrac{0.060}{30} = 0.002$ A

d $E = V \times I \times t = 1500 \times 0.0080 \times 10 = 120$ J

7. **a**

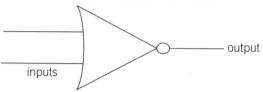

inputs

   b i low

   ii low

   c i

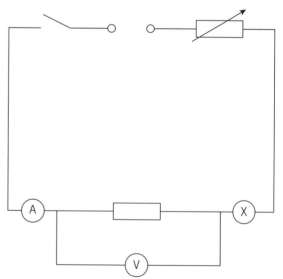

   ii No effect

Page 130

Practical question

1.

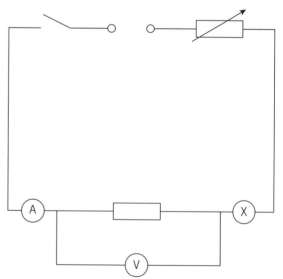

2. There is an ammeter reading the current.

3. Variable resistor.

4. Increase resistance of variable resistor; reduce voltage of power source.

5. A since current has to pass through more coils of wire.

Page 132

1. Loose live wire – earth wire
   Worn insulation – visual check of cables
   Wires get too hot – fuse or circuit breaker

Page 138

1. **a** **i** no current in circuit because the e.m.f. induced in AB is cancelled by e.m.f. induced in BC
   ii straighten out the wire
   b e.g. transformer, induction coil, generator, dynamo, microphone, alternator, computer

Page 139

2. **a** The wire has a magnetic field around it when a current flows through it; the field due to the magnetic poles interacts with this field and puts a force on the wire.
   b Out of the page.
   c Electrical to kinetic.
   d **i** commutator and brushes (see page 139 – d.c. motor)
   ii When the current flows through the coil a force acts on each side of the coil in opposite directions, due to the interaction with the magnetic field; the couple causes a turning effect on the coil; the commutator allows the polarity of the power supply to swap every half turn to keep the coil turning.

3. **a**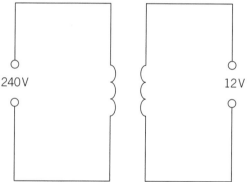

240V   12V

   b **i** For an e.m.f to be induced across the secondary coil, there must be a changing magnetic field in the primary coil.
   ii The changing magnetic field due to the ac current induces an e.m.f across the secondary coil.
   c **i** $P = I \times V = 1.5 \times 12 = 18$ W
   ii $E = 18 \times 3 = 54$ J

Page 140

4. **a** **i** 12
   ii The alternating current in the primary causes a change of flux linkage in the secondary coil, which induces an e.m.f; fewer coils in secondary so smaller e.m.f / voltage than in the primary
   iii heat in either coil /eddy currents in core / heat in core /magnetic leakage from core /sound from
   b **i** 12 V d.c.
   ii diode/rectifier
   c power in = power out, giving I = 8 A

5. **a** Attach the millivoltmeter across the wire; move the wire between the poles of the magnet.
   b The millivoltmeter would give a reading
   c generator

Pages 141–143

Summary questions for Unit 4

1. Charge – C; time period – s; current – A; frequency – Hz; potential difference or voltage – V; energy – J; resistance – Ω; power – W;
2. see table on page 121
3. **a** $Q = I \times t = 3 \times 6 = 18$ C
   b $I = \dfrac{Q}{t} = \dfrac{0.060}{100} = 60$ μC
   c $t = \dfrac{Q}{I} = \dfrac{15}{0.1} = 150$ s
4. $V = I \times R$
   a $V = 100 \times 0.5 = 50$ V
   b $I = \dfrac{V}{R} = \dfrac{10}{330} = 0.030$ A
   c $R = \dfrac{V}{I} = \dfrac{1.5}{0.8} = 1.9\ \Omega$
5. **a** $P = I \times V = 3 \times 12 = 36$ W
   b $I = \dfrac{P}{V} = \dfrac{40}{230} = 0.17$ A
   c $V = \dfrac{P}{I} = \dfrac{100}{0.5} = 200$ V
6. If the live wire touches the metal casing of an appliance, the current flows through the earth wire to ground; the current is large enough to melt the fuse and break the circuit, protecting the person from an electric shock.
7.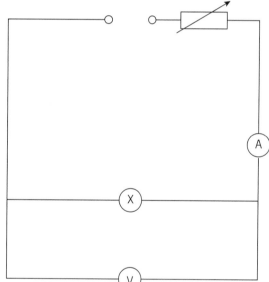
8. see page 125
9. A small current in the coil produces a magnetic field which attracts the iron armature; the iron armature pivots to close the contacts in the high voltage circuit, switching on a large pd.
10.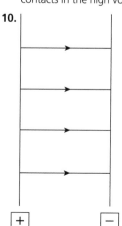

11. field; force; force; left hand; field; current; force; current; strength; magnetic field

12. see page 138

13. see page 134

14. see page 135

15. Spin the coil faster; stronger magnet; more turns on coil

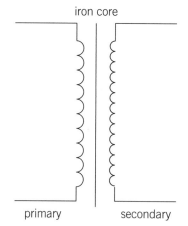

iron core

primary   secondary

16. The transformer requires a changing current to produce a changing magnetic field.

17. Across: thermistor, power, resistor, left hand rule, circuit breaker, ferromagnetism, induced emf, positive charge, transformer, resistance, south pole, transistor
Down: electric field, potential difference, relay, voltmeter, potentiometer, ammeter, current, not gate

Page 144

Examination style questions on Unit 4

1. a When the switch is closed a current flows in the circuit, through the metal spokes; the magnetic field due to this current and due to the fixed magnets interact and put a force on the spokes, which makes the wheel turn.

b Anticlockwise

d The commutator keeps electrical contact with the brushes but the brushes swap sides of the commutator every half turn. This keeps the coil moving in the same direction.

2. a $I = \frac{P}{V} = \frac{9}{6} = 1.5$ A

b i 8 Ω

ii 6 V

c i brightness decreases

ii Resistance in the circuit increases and so the current decreases, which reduces the brightness.

d i 4 Ω

ii 4 Ω

Page 145

3. a i $I_1 = I_2 + I_3$

ii $I_1 = I_4$

b i $V = IR = 0.8 \times 3 = 2.4$ V

ii $I = \frac{V}{R} = \frac{2.4}{2} = 1.2$ A

iii $R = \frac{V}{I} = \frac{(6-2.4)}{2} = 1.8$ A

4. a i Group 1

ii Group 2

iii Plastics

b

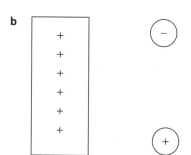

c Region around a changed particle where another changed particle experiences a force.

Page 146

5. a first finger – field / magnetism / flux; second finger – current / charge flow

b i brush / contact /sliding connector; split ring / commutator

ii clockwise

iii more current / more voltage; more turns on coil / more coils; stronger magnet; iron core

Page 147

6. a i Potential divider.

ii 12 Ω

iii $I = \frac{V}{R} = \frac{6}{12} = 0.5$ A

iv 5 V

v 5 V

b i 1:6 V, 2:0 V

ii

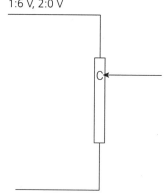

Page 151

1. Protons, electrons: 92; neutrons: 146

2. a 2 protons and 2 neutrons

b 1 electron

Page 153

1. a proton and neutron

b electron

c 4

d +2

2. alpha – into page; beta – out of page; gamma – not deflected

3. a An electron is removed from the atom by collision with the ionising radiation.

b Alpha radiation consists of helium nuclei, which are much more massive than beta radiation, which consists of electrons.

Page 155

1. **a** 0
   b −1
   c beta
2. **a** 4
   b 2
   c alpha
3. **a** 0
   b −1
   c beta

Page 158

1. **a** student C
   b half-life
   c **i** 4 hours
      ii 1
      iii 17 hours

Page 159

2. **b** Take count over one minute without the alpha source present i.e. the background count; take count over one minute with the alpha source very close to the detector; take a count over one minute with the source 3 cm from the detector; repeat each reading two more times.

Page 160

3. **a** alpha is unable to penetrate the cardboard box
   b gamma radiation would be unaffected by any mass of pineapple
   c Strontium 90 because it will not decrease in activity too quickly
   d too little juice: more; too much juice: less; no carton at all: more

Pages 161–162

Summary questions for Unit 5

1. alpha; beta; gamma; nucleus; gamma; alpha; paper; millimetres; deflected; positive; deflected right
2. electron – relative charge = −1 – relative mass = $\frac{1}{1800}$; proton – relative charge = +1 – relative mass = 1; neutron – relative charge = 0 – relative mass = 1.
3.
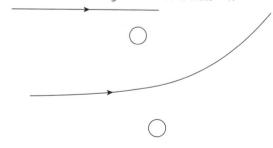

4.

Isotope	Number of protons in nucleus	Number of neutrons in nucleus
$^{14}_{6}C$	6	8
$^{3}_{2}He$	2	1
$^{238}_{92}U$	92	146
$^{231}_{95}Am$	95	136

5. Across: nucleus, paper, background count, plum pudding, uranium, alpha particle, lead, cancer
   Down: ionise, space, fission, cosmic rays, photographic film, beta particle, gamma ray, Geiger counter, Rutherford, half life, carbon

Page 163

Examination style questions on Unit 5

1. **b** β – deflected towards bottom of page, large deflection
      γ – not deflected
2. **a** 16 min
   b **i**
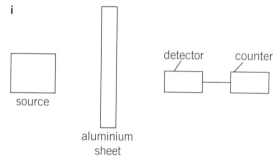
      ii Take count over 1 minute without the beta source; take count over 1 minute with beta source; take count over 1 minute with beta source and aluminium sheet in place; repeat each of the three readings twice more.
3. **a** 92
   b in orbitals around the nucleus
   c 146
   d in the nucleus
   e decreases by 3

Page 164

4. **a** **i** Time taken for half of the radioactive nuclei to decay.
      ii The radiation in the surroundings due to man-made and natural sources.
   b **i** 80 − 25 = 55 counts/minute
      ii 1:15 counts/minute, 2:40 counts/minute
      iii 28 minutes
      iv $\frac{25\ counts}{minute}$ = background count

Index